Marshaling Technology for Development

for Development

Proceedings of a Symposium

November 28-30, 1994
Arnold and Mabel Beckman Center
Irvine, California

Technology and Development Steering Committee

Office of International Affairs
National Research Council

Finance and Private Sector
Development
The World Bank

NATIONAL ACADEMY PRESS
Washington, D.C.   1995

NATIONAL ACADEMY PRESS • 2101 Constitution Avenue, N.W. • Washington, DC 20418

NOTICE: The project that is the subject of this report was approved by the Governing Board of the National Research Council, whose members are drawn from the councils of the National Academy of Sciences, the National Academy of Engineering, and the Institute of Medicine. The members of the committee responsible for the report were chosen for their special competence and with regard for appropriate balance.

This report has been reviewed by a group other than the authors according to procedures approved by a Report Review Committee consisting of members of the National Academy of Sciences, the National Academy of Engineering, and the Institute of Medicine.

This report has been prepared by the Technology and Development Steering Committee, which includes members from both the National Research Council and the World Bank. Support for the Symposium on Marshaling Technology for Development was provided jointly by both organizations.

A limited number of copies of this report are available from:

Office of International Affairs
National Research Council
2101 Constitution Avenue, N.W.
Washington, D.C. 20418

Additional copies are available for sale from:

National Academy Press
2101 Constitution Avenue, N.W.
Box 285
Washington, DC 20055
Tel: 1-800-624-6242 or (202) 334-3313 (in the Washington Metropolitan Area).

Library of Congress Catalog Card Number: 95-71477
International Standard Book Number: 0-309-05349-8

TECHNOLOGY AND DEVELOPMENT STEERING COMMITTEE

Co-chairs

Gerald P. Dinneen, Foreign Secretary, National Academy of Engineering
Jean-François Rischard, Vice President, Finance and Private Sector
    Development, World Bank

Committee Members

Jordan J. Baruch, President, Jordan J. Baruch Associates
George Bugliarello, Chancellor, Polytechnic University
Elkyn Chaparro, Senior Adviser, Finance and Private Sector Development,
    World Bank
Carl Dahlman, Manager, Private Sector Development Department, World Bank
Magdi R. Iskander, Director, Private Sector Development Department,
    World Bank
David P. Rall, Foreign Secretary, Institute of Medicine
F. Sherwood Rowland, Foreign Secretary, National Academy of Sciences
Vernon W. Ruttan, Regents Professor, University of Minnesota
Richard Stern, Director, Industry and Energy Department, World Bank
James B. Wyngaarden, Former Foreign Secretary, National Academy
    of Sciences

National Research Council Staff

E. William Colglazier, Executive Officer
Michael P. Greene, Program Director, Office of International Affairs (OIA)
Wendy D. White, Senior Staff Officer, OIA
Constance M. Reges, Program Assistant, OIA

iii

The World Bank, headed by its president Mr. James D. Wolfensohn, is a multilateral development institution whose purpose is to assist its developing member countries further their economic and social progress so that their people may live better and fuller lives. The term "World Bank" refers to two legally and financially distinct entities: the International Bank for Reconstruction and Development (IBRD) and the International Development Association (IDA). The IBRD and IDA have three related functions: to lend funds, to provide economic advice and technical assistance, and to serve as a catalyst to investment by others. Working with its borrowers, the Bank places poverty reduction at the center of its country assistance strategies.

The Vice Presidency for finance and Private Sector Development, led by Mr. Jean-François Rischard, was created in January 1993 to centralize the Bank's efforts to develop more vibrant and competitive private sectors in client countries. Most of the time of its approximately 200 specialized staff is spent helping the Bank's regional staff across all six regions to improve the quality and effectiveness of Bank assistance for private sector development, frequently through innovative lending and technical assistance operations. The remainder of their time is spent identifying and disseminating best practices in various areas, through training programs given in and outside the Bank, seminars and conferences, and multiple contracts with external partners of the Bank.

* * *

The National Academy of Sciences is a private, nonprofit, self-perpetuating society of distinguished scholars engaged in scientific and engineering research, dedicated to the furtherance of science and technology and to their use for the general welfare. Upon the authority of the charter granted to it by the Congress and signed by Abraham Lincoln in 1863, the Academy has a mandate that requires it to advise the federal government on scientific and technical matters. Dr. Bruce M. Alberts is president of the National Academy of Sciences.

The National Academy of Engineering was established in 1964, under the charter of the National Academy of Sciences, as a parallel organization of outstanding engineers. It is autonomous in its administration and in the selection of its members, sharing with the National Academy of Sciences the responsibility for advising the federal government. The National Academy of Engineering also sponsors engineering programs aimed at meeting national needs, encourages education and research, and recognizes the superior achievements of engineers. Dr. Harold Liebowitz is president of the National Academy of Engineering.

The Institute of Medicine was established in 1970 by the National Academy of Sciences to secure the services of eminent members of appropriate professions in the examination of policy matters pertaining to the health of the public. The Institute acts under the responsibility given to the National Academy of Sciences by its congressional charter to be an adviser to the federal government and, upon its own initiative, to identify issues of medical care, research, and education. Dr. Kenneth I. Shine is president of the Institute of Medicine.

The National Research Council was organized by the National Academy of Sciences in 1916 to associate the broad community of science and technology with the Academy's purposes of furthering knowledge and advising the federal government. Functioning in accordance with general policies determined by the Academy, the Council has become the principal operating agency of both the National Academy of Sciences and the National Academy of Engineering in providing services to the government, the public, and the scientific and engineering communities. The Council is administered jointly by both Academies and the Institute of Medicine. Dr. Bruce M. Alberts and Dr. Harold Liebowitz are chairman and vice chairman, respectively, of the National Research Council.

The Office of International Affairs (OIA) is concerned with the development of international and national policies to promote more effective application of science and technology to economic and social problems facing both industrialized and developing countries. OIA participates in international cooperative activities, engages in joint studies and projects with counterpart organizations, manages scientific exchange programs, and represents the Academy complex at many national and international meetings directed toward facilitating international cooperation in science and engineering. Dr. F. Sherwood Rowland, Dr. Harold Forsen, and Dr. David Rall are the foreign secretaries of the National Academy of Sciences, National Academy of Engineering, and Institute of Medicine, respectively.

iv

Preface

The last decades of the twentieth century have been characterized not only by a rapid pace of technological innovation, but also by the equally rapid integration of new technologies into society. The computer entered people's lives as a personal tool and transformed commerce and industry, from supermarkets to Wall Street. Other technologies propelled the telecommunications industries, providing the willing consumer with cable TV, mobile phones, and Internet.

These innovations are visible in any city of the industrialized world, and to a large extent in the newly industrialized states as well. But their deeper portent for the future is much less obvious, and their ultimate impact on the two-thirds of the world's population living in developing countries is the most difficult to assess. Yet for those responsible for the well-being of those countries—government leaders, businessmen and women, scientists and engineers—and for the development agencies, including the World Bank, that assist them, this question cannot be addressed too soon. If in fact the wave of technology innovation amounts to a second industrial revolution, as some predict, the developing countries must be prepared to join it or risk being left behind.

It is for this reason that the World Bank and the National Research Council proposed to combine their efforts to assess the impact of technological innovation on the developing countries. The collaboration that culminated in the Symposium on Marshaling Technology for Development was motivated by the sense that there are important challenges and opportunities for developing countries in recent technological advances, and that both development organizations and scientific institutions have a special responsibility to work together to assist these

v

countries in adapting to and profiting from these advances. The focus of this symposium was strictly on some of the new advances and the emerging technologies. Proven and well-understood technologies such as drainage systems, water recycling, and efficient stoves, whose value is undeniable, were not discussed; the same was true of nuclear power, which may be an important option for some countries but presently is not characterized as an innovative technology. Readers should be aware that this strict focus on the newer, advanced, and perhaps untried technologies is not meant to disparage older technologies by omission. On the other hand, whenever the observation is made that needed technologies are not available, it is based on an assessment of all the technologies, newer and older, available to the user.

The World Bank and the National Research Council have unqualified expertise on complementary aspects of the problem of applying technology to development. The Bank, as the world's premier development organization, has acted as lender to governments since the Bretton Woods agreement of 1945. The National Research Council (NRC), the operating arm of the National Academy of Sciences and National Academy of Engineering, is a nongovernmental scientific and technological organization whose primary function is to advise the U.S. government on issues related to science and technology.

Surprisingly, over the years the two organizations have interacted relatively little. In the early 1980s, they collaborated on a project to reform the Chinese universities in the fields of economics and engineering. A decade later, they joined forces in a project to assist the government of Indonesia to develop its capability to utilize technology in industry. In between, there has been only a handful of collaborative projects, most of them NRC studies that received partial funding from the Bank.

Separately, however, the two institutions have contributed much to the theory and practice of science and technology for development. For example, the World Bank has supported technology development through its lending operations, policy advice, and research, as well as through specific initiatives such as its support of the Consultative Group on International Agricultural Research (CGIAR) and its international network of research centers. The Bank's annual *World Development Reports* and monographs are well known for their reliable diagnostics of development problems and their influence on government policies. The National Research Council has produced many studies relevant to international development, including, within the last two years, *Realizing the Information Future, In Situ Bioremediation: When Does it Work? Vetiver Grass: A Thin Green Line against Erosion, Foundations of World Class Manufacturing,* and *Sustainable Agriculture and the Environment in the Humid Tropics.* Other NRC activities have directly supported research projects in developing countries and have convened technical workshops for U.S. and developing country scientists and engineers.

OBJECTIVES AND ORGANIZATION OF THE SYMPOSIUM

This collaboration was directed by a steering group, led by Jean-François Rischard, vice president for finance and private sector development of the World Bank, and Gerald P. Dinneen, foreign secretary of the National Academy of Engineering. The Symposium on Marshaling Technology for Development, convened November 28-30, 1994, at the National Academy of Sciences's Arnold and Mabel Beckman Center in Irvine, California, was designed to identify areas for action by the Bank and the National Research Council and to cement working relations among the technical staffs of the two organizations. It also served to initiate the formation of a network of scientific organizations on which the Bank may call for technical advice.

The symposium was formally opened by Bruce Alberts, president of the National Academy of Sciences, and Mr. Rischard, who introduced their organizations and briefly presented their personal views of science, technology, and international development. Keynote presentations were then made by George Bugliarello, chancellor of Polytechnic University, on the global generation, transmission, and diffusion of knowledge, and Harvey Brooks of Harvard University on technology transfer. These presentations were followed by three discussions of key generic technologies: biotechnology, led by Rita Colwell (University of Maryland Biotechnology Institute); materials, led by Praveen Chaudhari (IBM Corp.); and information technology, led by John S. Mayo (AT&T Bell Laboratories). A panel discussion chaired by Gerald Dinneen allowed these speakers to consider the broader, interdisciplinary questions proposed by the other participants.

The remainder of the symposium was devoted to sectoral sessions, each introduced by a primary speaker representing the National Research Council and followed by discussants from the World Bank and a developing country. The sectoral topics, selected by the World Bank, were areas of high technological content. The symposium did not pretend to cover topics not included in the sectoral sessions, nor to explore exhaustively the full range of views.

The primary speakers were Richard R. Harwood (Michigan State University), agriculture; Sidney F. Heath III (AT&T Corp.), manufacturing; Richard E. Balzhiser (Electric Power Research Institute), energy; Alan M. Lesgold (University of Pittsburgh), education and training; Jordan J. Baruch (Jordan J. Baruch Associates), services; Kenneth I. Shine, M.D. (Institute of Medicine), health; and Robert M. White (National Academy of Engineering), environment.

Discussants from the World Bank were Carlos Braga, Alexander McCalla, Mohan Munasinghe, Julian Schweitzer, Richard Stern, Peter Urban, and Jacques Van Der Gaag. Representing other institutions were Reynoldo dela Cruz, National Institute of Biotechnology, University of the Philippines; Agustin del Rio, Vitro Corporativo, Mexico; Demissie Habte, International Centre for Diarrheal Disease Research, Bangladesh; Lee Hoevel, AT&T Global Information Solutions; Vimla L. Patel, Centre for Medical Education, McGill University; Evald Emilevich Shpilrain, Institute for High Temperatures, Russian Academy of Sciences; N. Vaghul, Industrial Credit and Investment Corporation of India Limited; and James B. Wyngaarden, National Academy of Sciences.

Each sectoral session of the plenary was followed by a break-out session that included the primary speakers and discussants and other specialists from the World Bank and NRC. These sessions identified the important themes and formulated recommendations. Rapporteurs for the break-out sessions were Dennis Anderson, James Bond, Melvin Goldman, Kristin Hallberg, Lauritz Holm-Nielsen, Dean Jamison, and Christian Pieri from the World Bank; and Caroline Clarke, Donna Gerardi, Christopher Howson, Dev Mani, Susan Offutt, Procter Reid, and Paul Uhlir from the National Research Council.

This report on the findings of the symposium was drafted by Michael Greene of the National Research Council's Office of International Affairs and Kristin Hallberg of the Private Sector Development Department of the World Bank. It summarizes the ideas presented by the keynote speakers, the primary speakers, and the discussants, and incorporates the additions and recommendations formulated in the break-out sessions. The keynote and primary sectoral presentations appear in the third part of this report, Invited Papers.

ACKNOWLEDGMENTS

For the World Bank, Carl Dahlman and Elkyn Chaparro played key roles in the organization of the symposium. Their counterparts at the National Research Council were Michael Greene and Wendy White. Connie Reges managed the meeting logistics, assisted by Cindy Butler. The papers and proceedings were edited by Sabra Bissette Ledent, and Wendy White supervised the production process.

Contents

Appendix

Marshaling Technology for Development

INTRODUCTION

The Science of Sustainable Development

BRUCE ALBERTS
President, National Academy of Sciences

The Symposium on Marshaling Technology for Development was convened to initiate close collaboration between two great institutions on the problems facing the developing countries. Since 1945, the World Bank has been the world's premier development organization. The National Academy of Sciences and its associated entities—the National Academy of Engineering, the Institute of Medicine, and the National Research Council—are among the world's leading scientific organizations.

In 1863, under a charter granted by President Abraham Lincoln, 50 American scientists joined together to establish an American academy of sciences. The charter specified that the scientists were to advise the government on science and technology issues and receive no compensation for doing so. That tradition has endured—through the 1916 establishment of the National Research Council as the operating arm of the Academy and the later founding of the National Academy of Engineering and the Institute of Medicine, which are all part of the same "Academy complex."

Today, the National Research Council is a mature organization, operating with an annual budget of about $180 million. The three academies have about 4,000 members, and the National Research Council is staffed by 1,100 individuals, many of whom are widely experienced in science and science policy. Their work is to facilitate the efforts of the some 7,000 volunteer scientists and engineers who serve on about 600 committees, studying everything from what to do with the waste plutonium taken out of atomic weapons to what standards should be applied to the teaching of science in the public schools.

SUSTAINABLE DEVELOPMENT: AN INSTITUTIONAL PRIORITY

The topic of this symposium is of great importance to the National Research Council—in fact, the Governing Board of the National Research Council has designated sustainable development as a major area of focus over the next few years. To this end, the Academy recently initiated a Global Commons Project with a gift from a private donor, and a substantial part of income from the Academy's endowment fund is currently being used to support studies in this area.

But the topic of technology and development is, of course, bigger than one academy and one nation. One of the objectives of the Academy is to mobilize the world scientific community through the more than 60 existing academies of science and engineering. We aim to make a significant contribution to the application of science and technology to development, in part by bringing this important issue to the attention of the world's governments and institutions such as the World Bank. As a step in that direction, in October 1993 representatives of the world's science academies met together for the first time in New Delhi and drafted a statement on population that was presented to the United Nations International Conference on Population and Development, held in Cairo in September 1994.

Closer to home, in 1991 the Academy began to help Mexican scientists to create the same kind of relationship between their academies and their government that the National Academy of Sciences enjoys with the U.S. government—that is, the government asks the Academy to study specific problems; it funds the studies; and then it leaves the study committees free to produce their findings until the final products are released and announced to the world.

In 1995, the Mexican and U.S. academies of sciences and engineering carried out their first joint study, which dealt with the future of Mexico City's water supply. One of the byproducts of that study was a decision by the Mexican government to help set up a national research council in Mexico following the U.S. model. Similar study activities are just beginning with the academies of India and China on population and land use, and with the academies (or equivalents) of Egypt, Israel, Jordan, and the Palestine Liberation Organization on regional water resource management.

THE POWER AND PRODUCTIVITY OF SCIENCE

A look at the nature of science itself may help to answer the question: How can science contribute more effectively to international development?

Today, because governments are questioning their level of support for science, scientists have to try harder than ever before to explain to the public why they do science and how it is productive. People who have studied the productivity of fundamental scientific research have concluded that it is enormously beneficial to the economy and to society. In fact, recent analyses have found that the annual social rate of return from fundamental scientific research is between 20 and 50 percent.

Why is science such a powerful and productive endeavor? An analogy can be drawn from my 30 years as a cellular and molecular biologist. Unexpectedly powerful "system properties" emerge from what might be called organized complexity. The human brain is a prime example. Out of the billions of synapses between brain cells arise properties—"intelligence," "consciousness," and "will"—that cannot be deduced from the properties or actions of each individual brain cell. It is amazing that such a thing as our brain exists, and that it came about by natural selection during the evolution of complex human beings.

In biology, the complexity just described has been organized through trial and error—mutation, natural selection, mutation, natural selection—occurring for billions of years. In society, and in science, the role of the organizer of complexity is played by human intelligence. Consider, for example, the computer and telecommunications revolution. No one could have imagined how all the pieces of knowledge about electronics, materials, and mathematics would fit together and keep on developing in ever more powerful ways. In the end, what has been produced is something that is much different than anyone expected. And this has been characteristic of science and technology throughout history. Anyone who looks back now at the predictions made 30 years ago or even 10 years ago would recognize how wrong those predictions were. We must assume that the same will someday be said of our predictions, and we must keep our minds completely open to the impressive products that will be derived from science and technology throughout an unpredictable future.

All this leads back to the initial question: How can science contribute more effectively to international development? The more we recognize the power of science and technology and what it can do for society in practical ways, the more we must recognize that the world in the future is going to be a very different place. Advanced countries like the United States will presumably be enjoying all the benefits of science and technology. But other countries may have no significant scientific capacity at all. Moreover, the gap between the two types of societies could grow wider and wider. The old view was that science was a luxury for a developing country—an intellectual activity perhaps like a symphony orchestra, not an important priority. Such a view, however, is completely wrong.

CONDITIONS FOR THE APPLICATION OF SCIENCE AND TECHNOLOGY TO DEVELOPMENT

What are the minimum conditions that will allow a developing country to use science and technology to make the right kind of decisions and to prosper? The first requirement is the presence of science and technology expertise in each country and in each region. Scientists and engineers are multipurpose resources;

not only do they carry out research and development, but they also serve as interpreters and communicators of knowledge. They know what the needs are in their countries, and they can advise their own governments and their own industries of opportunities and perils in new technologies. They also can communicate with the world science and technology community to help to focus attention on the urgent problems of their region.

This symposium faces an important challenge: to define exactly the kind of international network of scientists and engineers needed to maximize the impact of science and technology on development. One example of the kind of new tools we should be using is a set of carefully designed data and discussion platforms on the Internet. Each platform would be designed to connect the scientists and engineers in developed and developing countries in a particular field, such as agriculture, water science, or health science. The platforms would have to be divided into sensible subcompartments—water purification, arid land irrigation, and so on—that are easy to use, even by novices. Each platform would include a database that provides access to the most useful recent review articles, as well as to abstracts of the literature already compiled in other electronic databases. A crucial feature of such a platform would be its facilitation of connections by electronic mail on Internet to a large collection of the appropriate human brains. Such a device would enable scientists in the developing world to connect electronically with other scientists who have the knowledge and experience applicable to the specific problem that they are confronting, whether these scientists are in a developed or developing country.

Today, not all parts of the world have access to the Internet. But they will have it soon; satellites and other devices will make it possible for a scientist to connect to the Internet no matter where he or she is in the developing world. This is an exciting tool that promises to alter the meaning of the term *international science*—providing that wise decisions are made now in planning for the future.

SUMMARY

As it evolves, science will become an ever more productive endeavor which will generate changes in society and in the world economy that cannot be predicted. The products of science and technology will be a crucial element in achieving "sustainable" economic prosperity. Capabilities in science and technology, however, are not well distributed among the countries of the world, but tend to be concentrated in the most developed countries. A very important priority for the future, then, is the improvement of such capabilities everywhere, building the capacity of all countries to use science and technology for their own prosperity. The communications revolution, in particular the Internet, will make possible new connections between scientists and engineers throughout the world, thereby creating the opportunity for a very powerful international science effort. The world's scientific academies and other scientific institutions—individually, collectively, and globally—will represent an important avenue for creating the many changes required to marshal the world's science and technology expertise to help create a more prosperous and sustainable world.

Forces Reshaping the World Economy

JEAN-FRANÇOIS RISCHARD
Vice President, Finance and Private Sector Development,
World Bank

The world has entered a period of massive shifts in its economy. Among the many changes likely to occur, China will be the world's largest economy by 2020; digital television and telephone systems will completely change the way people and businesses communicate; and such traditional activities as deposit banking may become shadows of their former selves. Behind these changes and the reshaping of the world economy are two major forces: a technology revolution and an economic revolution.

TECHNOLOGY REVOLUTION

A cluster of innovations centered around telecommunications and informatics has produced a revolution in information technology. Although this revolution is still young, with the full blooming of digitalization and of full bandwidth exploitation still some years away, it is a very powerful one. Three reasons are behind this potency.

First, information technologies are helping to unleash the potential of other technologies, creating subrevolutions in other areas. In transportation, for example, the container revolution, associated with hub airports, low-cost cargoes, and fast shipment methods, owes its existence to advances in information technology. Likewise, startling changes in high-performance materials, biotechnology, and robotics, and in programming and software applications of every possible kind, have been made possible by the information technology revolution.

Second, these information technologies are about information flowing faster, more generously, and less expensively throughout the planet. As a result, knowledge is becoming an important factor in the economy, more important than raw materials, capital, labor, or exchange rates.

And third, whereas earlier technology revolutions dealt with the transformation of matter or energy, the information technology revolution is all about time and distance. It is no great surprise, then, that the technology revolution is producing a companion revolution in business practices worldwide which can be illustrated by five examples.

First, evident almost everywhere today are accelerated, flexible business processes. Toyota, for example, has saved billions by adopting just-in-time inventory methods. Clothing retailer Benetton has a complete reorder cycle of two to three weeks—the time that elapses between purchase of a pullover in New York and the shipment of a replacement pullover manufactured in Sri Lanka. This was unheard of several years ago.

A second aspect of new business practices is hypercompetitive purchasing worldwide. Big U.S. department stores now solicit bids for cotton goods from 10 countries at a time. This was never the case before. Ford Motor Company is reorganizing itself around the concept of the global car, for which parts purchasing will be effected worldwide on a best-price basis. And electronic shopping networks are popping up, using electronic interactive catalog systems.

Third, smaller units—with smaller sizes, lower overheads, shorter feedback loops—are gaining the advantage over larger units. This trend will probably lead to the reemergence of family firms and, in general, the emergence of healthy, export-oriented, mid-sized firms. Big companies such as General Electric are breaking themselves up into a collection of smaller enterprises in order to maintain that small enterprise spirit. And flatter organizations--that is, with fewer hierarchical levels—are in vogue. For example, ABB, a large company that produces electrical turbines, just recently reduced the number of layers between its top management and the ranks.

A fourth aspect of this business practice revolution is the impending explosion of remote services that once were considered untradable. SwissAir has its revenue accounting done in Delhi. The Indian software industry already has captured a $500 million piece of the total turnover of the industry. In Washington, some doctors dictate into a telephone memos, which are then typed in Bangalore, India, in real time onto the doctors' computers in Washington. Even Romania, long backward economically, already has scores of teleporting firms in place.

Finally, there is the incredible flow of private capital—$175 billion last year—into developing countries, pouring in very quickly without regard for boundaries. (In the late 1800s, private capital flowed worldwide in this way.) As a result, the scrutiny of the World Bank or the International Monetary Fund (IMF) is being complemented by the scrutiny of private investors, who also are strong disciplinarians when it comes to demanding good management, open books, and disclosure.

ECONOMIC REVOLUTION

In addition to the technology revolution, with all the changes it has meant for business practices, there is the economic revolution, which has seen the massive entry of large new players into the world economy. When the IMF and the World Bank recalculated in 1993 the gross domestic products (GDPs) of all countries using the purchasing power parity method (it is better than current exchange rates), they were surprised to find that the non-member countries of the Organization for Economic Cooperation and Development (OECD) accounted not for a quarter of world GDP but approximately 46-47 percent, rising probably to about 50 percent if the black market economies are included. Moreover, in years past similar growth rates were calculated for both OECD and non-OECD countries. Now, however, the non-rich countries have growth rates that are two to three percentage points higher than those of the OECD countries because they have discovered capitalism and market-oriented policies. They also start from a lower base and have younger populations than the fiscally challenged, aging OECD countries.

When these two facts are put together, it means that two-thirds of the increases in world GDP will come from the non-OECD countries—the non-rich developing and transition countries—indicating an enormous shift in business opportunities toward the South and the East. Within this equation, China will regain its number one position in 2020, 200 years after it lost the title; India is poised for growth; Latin America will fare well despite the Mexican setback; Poland had the highest growth rate in Europe last year; and even some isolated countries in Africa and the Middle East are doing quite well. It has even been calculated that in 2010 the middle class in Asia—people earning $11,000-$12,000 a year—will comprise 750 million people. The twenty-first century, then, will see a replay for the developing world of what the 1950s and the 1960s were for Europe and Japan. That is what the economic revolution is all about.

A GOLDEN AGE?

Together, the technology revolution and the economic revolution are producing a completely new world economy that is high-speed, knowledge-driven, global, and disciplinarian. For 4 billion of the world's people it could be the birth of a Golden Age of sorts; for the first time they will have a serious opportunity to catch up with the rest of the world. They may even be able to leapfrog ahead in some areas. The planet, therefore, will become more balanced than it is today—a time when 15 percent of the people are consuming 85 percent of the goods and services. Another reason to believe in some kind of Golden Age is that the information technology revolution is likely to be followed in a generation or so by revolutions in biotechnology and solar energy.

It will, however, be a very stressful Golden Age because of tremendous demographic stresses as the world's population doubles over the next two generations. Environmental stresses in the form of deforestation, soil erosion, water and air pollution, loss of fisheries, and toxic wastes will be a serious concern everywhere. Much of the pollution will stem from the use of fossil fuels to generate energy. In the future, for example, India and China will account for two-thirds of the increase in world energy consumption. To meet the demand of its people, China will have to build one 1,000-megawatt power plant a month for the next 30 years, and most of them will be coal-based.

But to demographic and environmental stresses must be added a third kind of stress: the race for competitiveness—and the finish line keeps moving farther and farther away. To compete in this race, rich and poor countries alike need three things: agility, networking, and learning.

Agility is the *leitmotif* of this age, for firms as well as for enterprises. At the firm level, some U.S. shirt companies have reportedly returned 20 percent of their shirt production to the United States from the Far East because the turnaround time of U.S. shirt manufacture is so much shorter. These companies found that factor more important than the labor differentials with the Far East. An example of government agility: Singapore's customs system has reduced the time required to clear a ship through customs to 10 minutes by using electronic data interchange and other methods. Thus Singapore has now set the benchmark for agility in this area, which must be matched by everyone else in the world.

Networking, or getting plugged into global webs of relationships, is something all countries will have to work on, and there are many tell-tale signs that this is already happening. For example, strategic alliances in all their forms have tripled since 1990. Another sign of the age is the emergence of networkers in the form of big, cosmopolitan tribes such as the Sindhis in India and the overseas Chinese.[1]

Finally, countries will have to turn into learning nations because in the new global economy static comparative advantages are not enough; it is important to continually upgrade, learn, and stay with the flow. No one knows this better than the Colombian flower industry. After managing, over 10 years, to build a successful—indeed, miraculous—export business that was selling half a billion dollars a year in cut flowers to the United States, the Colombians are now struggling to defend their market share and their profitability against Dutch and other exporters, who have done a better job of upgrading their flower species, conservation methods, and transportation methods.

THE CHALLENGES FACING DEVELOPING COUNTRIES AND THE DEVELOPMENT COMMUNITY

The developing countries face massive new challenges. Not only will they have to deal with worsening population, environmental, and social problems, but they also will have to meet the ever-rising competitiveness threshold, expressed in terms of agility, networking, and learning, if they want to grasp the unprecedented opportunities offered by the new world economy. All this adds up to a future in which the distinction will be not just between rich and poor countries, but between fast and slow countries, plugged-in and isolated countries, learning and static countries. Thus the development job is not at all over; it is just becoming more complicated. The new development paradigm calls for vigorous action on three fronts: on the people and poverty front, on the environmental front, and on the growth and competitiveness front through private sector development, but also more generally through a kind of economy-wide pursuit of these higher agility, networking, and learning standards.

Now 50 years old, the World Bank Group, which includes the International Bank for Reconstruction and Development (IBRD), the International Development Association (IDA), the International Finance Corporation (IFC), and the Multilateral Investment Guarantee Agency (MIGA), is increasingly organizing itself and its agenda along the three dimensions—people and poverty, environment, and growth and competitiveness—to face the challenges of the new world economy. The institution itself is undergoing major changes. Since World War II, the World Bank Group has financed some 5,000 projects in 140 countries for a total of $300 billion, all based on less than $10 billion of shareholders' capital paid into the IBRD, which is the Group's main vehicle. The shareholders now number 178—near universality. Lending, however, has stagnated over the last five years, in part because it is being replaced by private flows. Meanwhile, the Group is becoming more and more involved in nonfinancial services—as objective advisers to governments, as best practice collectors and disseminators, as brokers, and even as fiduciary agents. Thus, just as growth and development are becoming more and more knowledge-intensive and less resource-intensive, the World Bank Group is moving beyond its money role and assuming, in addition, the role of purveyor of ideas and knowledge as it goes ahead. And it must do this with an even more diversified set of clients than before. The Eastern European countries, for example, have problems that differ drastically from those in Africa.

HOW CAN TECHNOLOGY HELP?

Today, technology probably has a bigger role to play in developing countries than at any time. Poverty can be tackled with the new technologies available in the fields of health care, population planning, basic education, food and agriculture, and infrastructure and basic services. For environmental problems, not only are there new technologies, but even existing ones will make a big difference in, for example, energy efficiency, exploration of new forms of energy, better forms of transportation, and new methods of fertilization. As for growth and competitiveness, the advanced telecommunications and informatics technologies will permit leapfrogging, as will the new distance education methods, production processes, and teleporting.

But how does one bring these technologies to bear on development problems? First, the developing countries must raise their leaders' and their populations' awareness of the imperatives of the new world economy and of the unprecedented role of new technological opportunities. Second, they must create an environment that is receptive to new technologies and to innovation. This will require:

- Ensuring macroeconomic, legal, and political stability and predictability
- Adopting policies that are receptive to everything that is open—open trade, foreign direct investment, foreign licensing, joint ventures
- Encouraging support institutions of the smart kind, particularly technical schools, business schools, applied research labs that serve as lookout posts for what is happening elsewhere, technology or productivity centers, and standards and patent institutes
- Building adequate power, information, transport, and financial infrastructures
- Enacting private sector-friendly policies that help to bring about a vibrant home base for enterprises, and, most important, for clusters of related enterprises.

Finally, to bring technology to bear on development problems a developing country must bring in the know-how and successfully ensure its implementation—that is, its incorporation into local production, marketing, and service processes. But this will be a very complicated undertaking.

In contrast, it is no longer as important to create local capabilities in basic research as it was two decades ago. Many of today's new technologies are more easily available and more sociable—some might even say promiscuous—than in the past. For example, the knowledge and software tools for designing new circuits can be taught in a classroom with a CD-ROM or downloaded over a telephone line.

OBJECTIVES OF THIS SYMPOSIUM

The purpose of the Symposium on Marshaling Technology for Development is to find ways in which the National Academy of Sciences and the World Bank Group can complement each other for the benefit of the developing countries. In such a strategic alliance between these two institutions, the National Academy of Sciences, through its operating arm, the National Research Council, could contribute its knowledge of new technological developments, their potential impacts, and what is needed to implement them. The Academy also has access to a network of researchers and scientists around the world.

The World Bank Group could contribute to such an alliance its knowledge of developing country conditions and institutions, as well as its access to a worldwide network of government and nongovernment institutions, development agencies, and business associations. Working together, these two institutions could offer a dynamic combination able to make a significant contribution to the developing countries as they face the unprecedented opportunities and challenges of the new world economy.

The specific objectives of this symposium are very practical: to get a sense of the trends and the impacts of new technologies in all the sectors that are important; to create greater awareness of the opportunities, particularly for those in the development community who have been lagging behind; to explore possible roles for the various actors—governments, private sectors, development agencies, and scientific institutions; and to determine what intelligent initiatives could be mounted, either in a sector or across sectors. But one of the nicest outcomes of this symposium would be an ongoing fruitful relationship between the two sponsoring institutions, the World Bank Group and the National Academy of Sciences.

NOTE

1. For a fascinating description of this phenomenon, see Joel Klotkin, *Tribes: How Race, Religion and Identity Determine Success in the New Global Economy* (New York: Random House, 1992).

PROCEEDINGS

CHAPTER 1

The Globalization of Knowledge
and Technology

In the last decades of the twentieth century, knowledge has become an organizing force in Western society, much in the same way that energy drove the industrial revolution. Knowledge takes the form of the fundamental science that underlies the new technologies transforming industry and commerce. It includes the data, news, and information generated at prodigious rates by firms, governments, universities, and international organizations. It also encompasses the timely information about trade, standards, prices, and business opportunities necessary for participation in competitive markets. Indeed, entire industries have been created to transmit, store, and organize knowledge and information, or to produce the devices that do. Computer technology, a tool only some 50 years old, is a vital part of these industries, and it also is affecting the everyday lives of a large fraction of the world's citizens through their communications, banking, health care, and workplace. Science itself has become an enormous enterprise, with billions of dollars invested in research and development worldwide, and research is constantly generating new knowledge that may be vital to human survival and prosperity.

In response to these developments, every country is challenged to prepare a strategy to generate, evaluate, disseminate, and act on the knowledge that is

This chapter draws substantively on the invited papers by Baruch (technological innovation and services), Brooks (technology transfer), Bugliarello (generation, transmission, and diffusion of knowledge), Chaudhari (materials and critical technologies), Colwell (biotechnology), and Mayo (information technology), as well as the discussions of the break-out groups. These four chapters, summarizing the findings of the symposium, were drafted by Michael Greene of the National Research Council's Office of International Affairs and Kristin Hallberg of the Private Sector Development Department of the World Bank.

needed today. Presently, many international, regional, national, and private entities are dedicating themselves to this question. For example, the Internet was created by the U.S. National Science Foundation and the Advanced Research Projects Agency (ARPA) of the U.S. Department of Defense precisely to provide a means of disseminating scientific and technical knowledge. With the additional efforts of thousands of universities and private entities, it has become the single most effective medium connecting scientists, engineers, and, increasingly, ordinary people in the world, often at no (perceived) direct cost to the user. The World Wide Web, which appears to represent the next generation of global knowledge resources, was a creation of the European Organization for Nuclear Research (CERN). It too is available to anyone with a computer and a modem, with no direct charge for the documents available on the system, most of which have been prepared by universities and private sources. A series of meetings of the heads of the industrialized countries is planned to coordinate development and expansion of these networks.

A second and perhaps greater challenge concerns the remedial role of knowledge and technology in ensuring continued human survival. Two hundred years ago, the Reverend Thomas Malthus noted that populations growing without constraint tended to increase exponentially, and he predicted that in a few generations the human population would exceed and exhaust the available food supply. That this has not yet happened is largely the result of a package of technologies known as the green revolution which has increased food production beyond any predictions, and of other technologies that have extended, found substitutes for, or protected threatened resources. Even so, hundreds of millions of people are malnourished, forests and per capita arable land are declining annually, and the world's population continues to follow an exponential curve. It is expected to double in about another four decades, and no one yet knows how food production will keep pace this time, how jobs and services will be provided, how territorial wars will be avoided, and how the environment will be protected. Also unknown

---

If the human race is to have a future, the global improvement of economic and social conditions through better knowledge is imperative. This is a problem that both developing and developed countries must address jointly. It simply does the world no good in the long run if individual countries succeed in addressing their socioeconomic problems at the cost of neglecting such global problems as the growing depletion of the ecosphere or the potential for international conflict.

—GEORGE BUGLIARELLO

is whether the resources for a growing population will always be found in the future, whether population can be contained—possibly through a combination of a higher standard of living, new contraceptive technologies, and social changes—or whether the cycle of crisis and technical fix will repeat itself until finally the fixes can no longer keep pace. Ultimately, of course, human population growth will cease.

PATHWAYS TO TECHNOLOGICAL INNOVATION

For developed and developing countries alike, a country's ability to realize gains in knowledge-based productivity depends on its capacity to tap the global system of generation and transmission of knowledge through technology transfer, generate indigenous knowledge through research and development, put that knowledge to productive use through engineering, ensure equitable and effective use of that knowledge through social and behavioral research, and organize and diffuse information. Technology transfer—the mechanism for bringing a technology from one research area of industry to another or from an advanced to a developing country and putting it into operation in its new environment—and research, development, and engineering—the process of conceiving or adopting a new idea, developing a technology, and adapting it for practical use—are generally considered separate and distinct processes. In reality, however, they are interconnected. There is a much smaller difference than is generally supposed between introducing a technology that is new to the world and one that is merely new in a particular sociotechnical context characterizing a particular manufacturing site and market. For this reason, the process of creating a production system at a new site can be considered, at least in part, an innovation. The process is fundamentally similar in developed and developing countries, whether one is replicating something that has been done before elsewhere or doing something that has never been done before anywhere. It is the absorption of knowledge into a system of product or process realization involving design, production, and marketing to clients and customers that is critical. Yet beginning with the knowledge that the technology has "worked" elsewhere lends a significant advantage.

This view is reinforced by the finding that even in the developed world research and development represent only a small fraction—10-15 percent—of the resources required to bring to market a new product incorporating substantially new technology. The other 85-90 percent is so-called downstream investment—design, manufacturing, applications engineering, and human resource development. In fact, about 65 percent of the scientists and engineers in the U.S. national work force are not even engaged in research and development, but in a broad spectrum of activities related to these downstream elements. The same is true in most developing countries, where scientists and engineers carry out little research and development, and the downstream elements represent an even larger fraction of the total effort.

One of the important trends of the last 30 years in the developed countries has been the increased importance of sources of technical information and ideas external to the producer, including institutional alliances and "innovation networks" (links between users and producers), some of which cross national boundaries. The result has been a complex intermingling of competition and cooperation, with some firms cooperating in selected projects while at the same time competing in others.

These alliances also have brought into relief the roles of the two faces of technology development, the supply side and the demand side or, put most simply, "solutions in search of problems" and "problems in search of solutions." Solutions in search of problems relate to fundamental research and development, which are the basis of much of our understanding. Problems in search of solutions are what industry, society, and design engineers encounter in practice. The process of technological innovation matches solutions in search of problems, whether found in the laboratory or the library, to problems in search of solutions. Development activities, too, can yield important new technologies. The closer these activities are to applications, the more productive they may be in the near term. The proportion of resources dedicated to each type of activity depends on the technical capacity of a country and the level of fundamental knowledge already available on the problems of interest. In agriculture and in health, the problems may be inherently local, requiring substantial fundamental research to produce enough knowledge to support applied research and development on specific problems. In industrial technology, the proportion might be changed. Japan in the twentieth century and the United States in the nineteenth century reached a high level of technological development with a minimum of fundamental research; the United States, adopting a different strategy in the twentieth century by increasing its investment in fundamental research, became the world's leader in both science and technology.

Another important distinction in technology development is between radical and incremental innovation. *Radical innovation* produces new ways of doing things and ultimately leads to new services or industries—the computer and the flat-screen display are good examples of such innovation. *Incremental innovation* is a term applied primarily to relatively small improvements in existing products and processes or to relatively small extensions of the scope of existing applications of the product design or technology, which over the long term accumulate to produce major changes. Most technological innovation is of this type, and much of it takes place on the factory floor (or clinic or farm) by the users of technology. In fact, steady productivity growth and the expansion of both the size and technological scope of markets primarily stem from the cumulative effects of many apparently minor incremental innovations. Because such innovations are just as important to economic success in developing countries as in developed countries, it is important that Third World firms build their technological capacity and engineering capability.

This being said, why is radical innovation a concern at all of developing countries since most such innovations are created in the industrialized countries that possess the necessary science and technology infrastructure? The answer is that both types of innovation, the radical and the incremental, generate new business opportunities that do not necessarily require the same level of advanced knowledge and capacity that was needed to create the radical innovation in the first place. Developing countries with a minimum level of basic education in their work force and minimum industrial experience may be able to capitalize on these opportunities quite successfully, often at lower cost than developed countries. This has been demonstrated many times by the success of several of the newly industrialized countries in finding niches in the computer and information technology markets.

In all the niche-type opportunities . . . , the aspiring entrant must have a fairly thorough understanding of the technological system in which a potential new niche may lie. This is one reason why imitation can be said to be the first step toward innovation— but only to the extent that it provides a real window into an entire technological system.

—HARVEY BROOKS

Many of the radical innovations that have the potential to reshape the world originated in scientific advances made over the last few decades. Some of the most important advances from the point of view of international development have been in three specific scientific and technological fields: information and communications, biotechnology, and materials science. Advances in these fields have led to technologies that are basic to many different products.

Information and Communications

The rapid rate of progress in information and communications technology over the past few decades has been called a revolution. The key technologies underlying this revolution are computing, fiber-optics communications, software, and silicon chips. Progress in computer technology is measured by processing speed and memory, progress in fiber optics, and cost per unit of bandwidth. The costs of computer memory and bandwidth have dropped by a factor of two over the last five years. Even software—once considered a bottleneck technology because of quality problems—is beginning to advance rapidly in such major areas as telecommunications, thanks to advanced programming languages and reuse of previously developed software modules instead of writing and debugging new ones.

The power of silicon chips is measured by the number of transistors that can be placed on a chip; a typical chip measures about 1 square centimeter. Each time the number of transistors on a chip has increased by a factor of a thousand, integrated circuit functions have radically advanced. The earliest chips held only one transistor. When research and development placed 1,000 transistors on a chip, engineers were able to replace analog with more powerful digital circuitry. About a decade ago, the number of transistors reached 1 million, enabling microcomputers to perform functions that approximated the mainframes of only a few years earlier. The present goal is 1 billion transistors per chip, and some people in the industry predict that will lead to another revolution: the merger of communications, computers, consumer electronics, and entertainment.

For the greater part of this century, the telecommunications industry and information highways were paced by the availability of new technologies. But today, with a wide array of multimedia and information technologies available, the telecommunications industry is being driven increasingly by customer demand, leaving many possible innovations unexploited. And in another important trend, the global transfer and assimilation of information technology are combining with political, economic, and regulatory forces to produce, for example, the move toward privatization of telecommunications in both developed and developing countries. The result is increased global competition in the provision of communications products and services, which should result in lower prices, new products, and response to market pressures, as occurred in the computer industry.

At the same time, however, there is a worldwide push for common standards and open, user-friendly interfaces that encourage global networking and maximum interoperability and connectivity. The evolving international standard for photonics or lightwave transmission devices such as optical fibers is called synchronous digital hierarchy or SDH. It will enable users to purchase equipment from many different vendors without worrying about compatibility, and it will

---

*This reengineering of the communications industry
appears to be the next to the last step in the information
revolution brought on by the invention of the transistor. The last
step, and one that may go on forever, is the reengineering of
society—of how people live, work, play, travel, and communicate—
creating a whole new way of life.*

—JOHN S. MAYO

allow vendors in developing countries to compete on an equal basis. But it may well be that the opportunities opened up by these developments will not fully appear until the present fluid situation, characterized by intense competition among large numbers of small suppliers and uncertainty about future directions affecting many different sectors, crystallizes into a mature industry.

Biotechnology

Biotechnology is defined by the U.S. government as "any technique that uses living organisms to make or modify products, to improve plants or animals, or to develop microorganisms for specific uses." It has been employed successfully for hundreds of years to manufacture medicines, to improve agricultural production, to produce drugs, and, in the form of fermentation, molds, and bacteria, to produce food products. Over the last two decades, however, a "new" biotechnology has been defined as the use of recombinant DNA and other genetic engineering techniques to produce new organisms and new products. This biotechnology has enabled researchers to accelerate the rate of innovation and apply new and effective techniques to new areas and new problems.

Genetic engineering was born in 1944 with Avery, MacLeod, and McCarty's paper revealing that DNA is the genetic material. This discovery was followed in 1953 by the landmark paper by Watson and Crick describing the helical structure of the DNA molecule. In 1973, the work of Cohen and Boyer showed how to transfer a gene from one species to another. Confirmation of genetic engineering as an industrial technique came only seven years later in the 1980 Supreme Court decision of *Diamond v. Chakrabarty*, which allowed microorganisms to be patented. In 1994, the 1,300 biotechnology companies located in the United States alone invested $7 billion in research and development. In addition, nearly 400 companies are located in Europe, 300 in Canada, and a few hundred more in the rest of the world.

A novel feature of the new biotechnology industry in many countries is the proliferation of small, entrepreneurial start-up companies. But many of these eventually fail or merge or are bought up by larger companies because of the unexpectedly slow progress and high cost of bringing products to market. This competitive situation may prove to be a benefit for the developing countries, where biotechnology companies have begun to appear, enabling them to form strategic alliances with more technologically capable foreign companies looking for new markets.

Biotechnology has a potential impact on many areas of agriculture, health, industry, and environmental protection. In the United States, biotechnology has found its greatest use in health and biomedical applications—mostly recombinant drugs, enzyme-mediated diagnostic kits, and designed pharmaceuticals. Some of the new products likely to affect developing countries are test kits for such infections as HIV and malaria and recombinant vaccines for some of the world's major diseases. Another promising application is gene therapy, which presently is extremely expensive but could eventually solve such problems as Parkinson's disease and sickle-cell anemia.

In the developing countries, biotechnology will likely be predominantly applied to agriculture in the form of transgenic plants, produced by combining the genetic material—and therefore the characteristics or nutritional benefits—of more than one species; biological pest control, in which the pesticide is produced directly by the crop plant itself; tissue culture for mass generation of desirable cultivars (and its counterpart in livestock, *in vitro* fertilization); disease prevention and control in crops and livestock; and new agroindustries for the production of fuel or industrial raw materials. At the same time, biotechnology in the hands of the industrialized countries will produce substitutes for commodities now providing income or subsistence for developing countries.

Marine biotechnology accounts for only a small fraction of the world biotechnology market, but it has a high potential for benefiting the developing countries. Although the oceans presently provide less than 1 percent of the world's food calories, the demand for seafood is expected to increase by about 70 percent in the next 40 years. This demand comes, however, at a time when many fisheries are declining because of overexploitation driven by new harvest technologies. To meet the increased demand while natural stocks are in decline, world aquaculture production would have to increase by more than seven times. Biotechnology could play a vital role in efforts to improve captive management, promote faster reproduction of species and the production of healthier organisms, and improve the food and nutritional qualities of the organisms, including the introduction of new "crops," such as algae and seaweeds, for their nutritional properties. It must be remembered, however, that aquaculture is energy-intensive by nature, and its efficiency will be related to the cost and availability of energy.

Among the environmental applications of biotechnology, bioremediation uses both naturally occurring and genetically engineered organisms to clean up polluted sites by transforming toxic and other undesirable materials into more benign or volatile substances. Bioremediation is applied, among other things, to oil spills, industrial wastes, soil contaminated with TNT or heavy metals, raw sewage, and polluted bodies of water. Test kits and sensors for environmental monitoring also are becoming available. This technology is demonstrated quite visibly in the main square of Jakarta, Indonesia, where a prominent electronic billboard indicates levels of atmospheric contaminants.

Mining, forestry, energy, and bioelectronics applications of biotechnology are expanding. Engineered microorganisms to remove ores may increase the efficiency of mineral extraction while reducing pollution at mine sites. Tissue culture will assist forest restoration, and the development of alternative biofuels will slow the destruction of the world's forests while providing an alternative to petroleum-based fuels. One area in which it is particularly hard to predict future developments is biologically based electronic components, including computer chips. Recent work indicates that biological systems might be designed to operate more efficiently and rapidly than semiconductors for some applications.

Materials

Over the last century, science has learned not only how to identify the properties and structure of materials existing in nature but also how to combine the atomic elements to produce artificial materials. Using a recently invented scanning microscope, scientists can image individual atoms on a surface and even move them, one by one, to new locations. This requires resolution and manipulation on a scale of less than a billionth of a meter, the order of the width of one atom. Similarly, scientists are now able to observe phenomena that span a millionth of a billionth of a second.

Great advances are being made in developing materials that have the desired thermal, optical, and mechanical properties, are easy and less costly to manufacture, and are durable or biodegradable. Some of the most promising are found in those areas where materials science touches informatics and biotechnology. Two examples are the magnetic resonance imaging (MRI) used in medical diagnosis and the "intelligent" materials being used in modern prosthetics. In these examples, the behavior of the materials is controlled by computer or chip so that they can interact with a living entity. These technologies, which are among the most sophisticated and expensive in use, have little application at the moment in most developing countries, mainly because of their costs, but they probably will be used widely in the next century. Other advanced materials are now finding application in developing countries as cost-effective replacements for earlier technologies—for example, the use of shape memory alloys for affordable automatic control, amorphous silicon solar panels for roof-top power supplies, and advanced magnetic materials for small motor actuators.

No major nation can avoid facing the questions associated with its semiconductor chip design and production. Their use will be pervasive—from feedback control in prosthetics to voice recognition devices used to control such mundane subsystems as locks in houses, radios, television, scooters, cars, and computers. Every human family, in some form or fashion, will own a silicon chip in the near future.

—P. CHAUDHARI

Also promising for developing countries are the technologies at the intersection of materials and informatics or communications. The steady evolution of the integrated circuit and silicon chip over the past few decades provides a measure of improvements in materials technology. The global importance of these technologies stems from their use in hundreds of applications and their falling costs and size over time. Even as the earth's human population grows, a future in which silicon chips are so cheap and ubiquitous that every family owns at least one is not improbable. Indeed, the size of the markets in number of units could exceed those for nearly every other manufactured commodity. A companion product is the flat display, a quickly evolving product employing a wide variety of alternative technologies. No one yet has dominance in this market, which could embrace one out of every four families within a decade.

Thus advances in materials science and technology, especially where combined with biotechnology or informatics, will lead to further radical innovations and will yield new classes of products that will have enormous markets in the near future. Countries not making these products will be buying them. And, like many radical innovations, these advances will create a large number of niches in the market for component and materials suppliers and technical services that may be filled by developing country firms positioned to compete.

TECHNOLOGICAL CHANGE AND THE GLOBAL ECONOMY

Even though much uncertainty still surrounds the technology revolution, one thing is highly probable: the speed of innovation in information and communications, as well as biotechnology and materials technologies, will reshape the world economy, creating new industries, changing the nature of markets and the sources of comparative advantage, lessening the importance of geographic boundaries, and changing the way business is done. Many developing countries will encounter new opportunities to increase their productivity, incomes, and participation in world trade. Yet at the same time, these same countries will have to adjust to the economic and social change brought about by the new technology. Countries that fail to adjust and use technology to their best advantage probably will fall behind those that do.

At the level of the individual firm, a knowledge-based global economy will put a premium on speed; a rapid pace of technological change means that current knowledge, including the knowledge embodied in human and physical capital, becomes quickly outdated. Workers may find that the knowledge learned in school or in early years on the job is not sufficient to deal with new advances. And as physical capital also becomes obsolete more quickly, firms failing to keep up with technological advances may find themselves lagging behind their competitors. Some say, in fact, that in the new global economy there will be two types of firms: the quick and the dead.

Goods previously thought to be nontradables, such as services, will become commodities via new information and communications technologies. Transborder services already are the fastest-growing component of both international trade and foreign investment, opening new export opportunities for developing countries. For example, many Caribbean countries are exporting services to the United States, including financial (check clearing, insurance claim processing), communications (toll-free line answering), and tourism (hotel reservations). In India, the software industry has been able to take advantage of its low-cost, highly skilled work force and international communications links to become a major exporter of software. Even the service components of manufacturing, once embodied in manufacturing production (for example, design, mapping and geological services, and accounting), can be outsourced. In fact, this unbundling of services, by increasing the measured domestic production and trade in services, largely explains the rising share of services in the U.S. economy.

In the changing global economy, the new technology will increase competition and contestability in markets by lowering barriers to entry, reducing the minimum efficient scale of production, and providing alternative production techniques. Information technology will increase the contestability of service industries by improving consumer access to information and by opening opportunities for long-distance services. Some estimates suggest that as much as 10 percent of the 88 million service jobs in the United States could be contested by long-distance suppliers under the right set of circumstances. Industries long considered natural monopolies, such as telephone services, already have become more competitive with the entry of new service providers using new technologies such as cellular telephones. Eventually, these new market structures will erode existing regulatory frameworks, making them ineffective, inefficient, or irrelevant.

As barriers to entry and scale economies are reduced or become ineffective, smaller-scale firms will find that they can compete with larger ones, potentially flattening the size distribution of firms in both national and international markets. Information technology can be used to level the playing field among market participants—for example, by reducing information asymmetries between buyers and sellers and eliminating the need for middlemen. An example of the latter: on-line catalogues are beginning to appear on the Internet, and the Home Shopping Network has been valued at over $1 billion. Another example of this trend toward exploiting the latest information and communication technologies is The Limited, a chain of clothing stores for fashion-conscious women in their twenties and early thirties. Using real-time information from its cash registers, this firm restocks its stores by automatically ordering garments from suppliers in China via a satellite communications link. As a result, the firm has dramatically reduced its turnaround times to about eight days, keeping inventory costs at a minimum and posing a challenge to its competitors. In industries like The Limited, then, the mass production of customized products has been made possible by technology. In this and other industries, firms that have not gone through the large-scale, mass-production phase of industrial development may find that their existing business structures give them a competitive advantage in external markets.

The world economy, therefore, depends greatly on technological change. Indeed, many of the leading industries today were unknown or vastly different a century ago. Also today, for the first time in history, the world is able to produce enough food to feed all its present inhabitants, even though for a variety of complex reasons pockets of hunger still linger. Effective health care is reaching populations that never before have benefited from modern medicine and is a major factor in the burgeoning population growth the world is experiencing. To encourage technological innovations, most industrialized countries have established national research and development institutions and devised systems or programs to incorporate innovations, whether domestically generated or not, into the national productive sector. Many developing countries, however, lack both. How then can these countries tap into this rich vein of technological change? This question is the subject of the next chapter.

CHAPTER 2

Marshaling Technology for Development: Assessing the Challenge

Technological change is a complex process. For some radical innovations—as was the case for the airplane, radio, and nuclear weapons—curiosity, new knowledge, or political will create a need; the need will lead to an invention; and the invention will evolve into a new technology. The new technology, if successful and useful, will diffuse and will replace older technologies, just as the railroads replaced canal transportation and motor vehicles and airplanes replaced railroads. It will then spread through trade, the exchange of information, deliberate technology transfer, conquest, or industrial espionage.

For incremental innovations, the process is more subtle, even imperceptible. Because of modifications on the factory floor, one firm's products will be better designed, more efficiently made, or more effectively marketed than those of other firms. Success will generate imitators, and the new product gradually will become the industry standard, pushing aside similar products and, if it continues to improve, replacing other technologies. This process is more analogous to the evolution of biological species, and it can be just as effective at bringing radical change.

CONDITIONS FOR CHANGE

But exactly what are the conditions that encourage invention and innovation, both radical and incremental, and facilitate the acquisition of technology? This

This chapter draws substantively on the invited papers by Balzhiser (energy technology), Bugliarello (generation, transmission, and diffusion of knowledge), Harwood (agricultural development), Lesgold (educational technology), Mayo (information technology), and White (sustainable development), as well as the discussions of the break-out groups.

29

question has been studied extensively at the firm level, but almost entirely in the developed countries and in the context of competition among developed countries. Less clear are what fundamental conditions within a developing country enable producers to be innovators in world markets.

A modern, effective technological infrastructure enables a country to generate and utilize knowledge and innovation. The evaluation and acquisition of new technologies are made easier when the society and the economy, including availability of investment capital, are intrinsically receptive to innovation. The important elements of such a technological infrastructure are:

• An educational system that encourages creativity and the pursuit of scientific and technological knowledge
• An educated and skilled work force
• A network of capable research laboratories, linked together and able to gain access to scientific and technological information from the outside world
• Facilities for product development and quality control, including testing and standards laboratories responding to international standards
• Critical technical resources, including machine shops, precision foundries, and computational facilities
• An industrial structure that will sustain a productive "industrial ecology," in which small, technically oriented and potentially innovative suppliers serve larger firms that have access to markets and resources
• Institutions or programs that link researchers and inventors to the potential users of the knowledge they generate, as well as to investors
• A legal system to protect technological innovation, whether indigenous or imported
• An economic policy environment that encourages research and development and investment in innovation
• A reliable electric power network with good frequency and amplitude control
• An adequate transport, communications, and telecommunications infrastructure.

For successful innovation, it also is imperative that a country's people and their leaders view the effective utilization of technology not just as an option but as the key to successful and sustainable development—and that they be prepared to act. One of the factors undoubtedly contributing to the economic success of Taiwan in recent years is the number of Ph.D.s in leadership positions, including the current cabinet members and the prime minister. Among many countries, especially the poorer ones, a widespread commitment to technology has yet to emerge, hindering the process of technological development that has been so successful in Europe, the United States, Japan, and other parts of Asia. For those developing countries, it is essential to emphasize technological literacy in the schools—indeed, among the entire population.

*Obstacles to the universal adoption of biotechnology
projects and products are cultural, educational, economic,
governmental, and infrastructural in nature. If, for example,
difficulties are encountered in delivering agricultural products
to market, no change in the qualities of those products will
overcome the infrastructural problems. In other words, there is
no reason to introduce genetically engineered apples that
ship better in a region where the apples rot on the trees
because they cannot be shipped to market.*

—RITA COLWELL

It also is important that each national research community concentrate on the needs of its country. For a developing country, this usually means a strong emphasis on applied research and development, working on problems in search of solutions, without neglecting the basic research that provides a base of talent and a window on the knowledge generated elsewhere. It also means having the flexibility to cut across the traditional academic barriers to carry out the interdisciplinary studies that are so important in the technological market.

Management issues are also critical. New technologies usually require new techniques for their management, starting with selection and acquisition of equipment, and proceeding to training of workers, process and product design, marketing, quality control, cost containment, labor utilization, and safety. Learning from the experience gained in other places is the best method, where possible, but attention must be paid to local conditions—economic, physical, and social—and local capabilities. Transfer of technology must embrace transfer or development of the skills necessary to manage technology.

Finally, developing countries will find it advantageous to pool their scientific, technological, and educational resources with those of other, similar countries because few of them have enough skilled scientists and engineers and resources to make many breakthroughs alone. They must be receptive to the importation of new technologies, whether through foreign direct investment, joint ventures, or licensing from other countries. The ability to share resources and identify opportunities for technology transfer will depend on the establishment and maintenance of communications, transport, and personal links among countries.

CONNECTIVITY

It is one thing to recognize that the information and technology desired are available, but it is quite another to gain access to them. That will require that the developing countries strengthen their linkages with the rest of the world by investing in the infrastructure needed to receive and transfer information. In this undertaking, partnerships are key: between research institutions in developed and developing countries, between domestic and foreign firms, and between research institutions and the private sector.

The physical infrastructure needed to support connectivity includes computers and telecommunications hardware and software (including connections to the Internet and World Wide Web) and reliable supporting power and communications networks, as well as the transportation basics—roads, airports, navigation. A modern information and communications infrastructure will provide up-to-date technical information and publications and will allow instant communication among scientists around the world. Technical information services linked to worldwide information networks can distribute knowledge quickly and cheaply to the productive sectors. Thanks to satellites, the developing world can leapfrog immediately to advanced telecommunications capabilities, bypassing the long road already traveled by the industrialized world.

Direct international links between firms are important vehicles of technology transfer. The observed correlation between exporting and growth in productivity arises partly because exporting is an extremely effective source of learning. Foreign direct investment provides not only capital but also technology, management, access to global networks of information, and access to markets. Strategic alliances—between foreign and domestic firms in joint ventures and between domestic firms in the same or related industries—create pathways for the transfer of knowledge. In addition, much technology is transferred through informal means—copying, reverse engineering, reading technical journals, attending foreign conferences and trade fairs, and hiring foreigners with technological expertise.

But how do the users of technology channel their feedback to the generators of knowledge? The establishment of linkages between networks of research

---

Regardless of the undeniable improvements in electronic and computer communications, the need for scientists working together to interact daily is unchanged, if only because science is increasingly cross-disciplinary and thus demanding of team effort. This is no less true in developing countries and may be even more urgent in some respects: efforts to develop and extend technologies in most such countries have often failed as a result of the lack of well-articulated teams of trained personnel to facilitate real technology transfer.

—KENNETH I. SHINE, M.D.

laboratories and the private sector will facilitate such feedback. Clusters of research-production-marketing activities, such as applied research parks or university-industry parks, would serve to better link research to production and the marketing of the results. Rather than dispersing assets, such parks can offer synergistic concentrations of knowledge workers and facilities.

In the agricultural sector, much technology transfer occurs horizontally—farmer to farmer. In industrialized countries, many farmers use electronic networks and other modern information technologies to make this process more effective. This would likely work in developing countries as well, thereby facilitating technology transfer among farmers and strengthening their linkages with markets and price information. In the education sector, electronic networking could put teachers in contact with one another, provide classroom materials from domestic or international sources, as well as give both teachers and students hands-on experience in using communications technology.

THE NEGATIVE IMPACTS OF TECHNOLOGY

The introduction of new technologies may provide beneficial economic opportunities or a necessary lifeline for feeding, employing, and ultimately limiting an expanding population. But negative impacts on society will be felt as well, bringing environmental, social, and political change. If some of these effects of new technologies can be predicted, steps can be taken to recognize and prepare for them.

Technological innovations often make critical industrial or agricultural products uncompetitive or obsolete. This happens in developed countries, as some American textile producers well know, and will certainly occur in developing countries. Such an effect will put at risk any country that fails to anticipate change, or whose economy relies heavily on a few traditional products. Likewise, in those countries where manufacturing is an important source of employment for the marginal or uneducated worker—the case in much of the world—the introduction of manufacturing practices that are more knowledge-based and specialized will cause dislocations, presenting difficult educational and political challenge in most countries. Attempts to hold change at bay, however, by closing borders to innovation will be counterproductive even in the short term. The restructuring of industry and the retraining of workers will require major decisions, significant resources, and, where urgent, outside assistance for the affected communities.

Exposure to new technologies and to international culture and fashions may weaken long-standing traditions and practices, especially among the younger generations. This effect occurs everywhere, but it is less disruptive when the changes come about gradually from indigenous developments. Developing countries may need to fortify key elements of their culture through special school programs and by celebrating and encouraging local literature, music, and preservation of historical sites and local languages. Sometimes the new technologies themselves will assist in the preservation, enjoyment, and even exportation of the local culture. One country's IMAX productions featuring its diverse cultures and natural beauty are seen each year by thousands of school children locally and help to attract many tourists from abroad.

Over the long term, there is potential for the convergence of developed and developing economies through universal access to knowledge. Over the short term, however, the capacity to benefit from access to knowledge, technologies, and markets will depend on the initial stock of educated human capital and real capital. Productivity in services and information-intensive activities will grow faster in the North than in the South, and faster in the newly industrializing countries than in the least developed. Accordingly, further economic divergence may be the outcome in the initial stages of the process.

The danger of a widening gap between technology haves and have-nots exists not only across countries or groups of countries, but also across socioeconomic groups within individual countries. Equity issues surrounding different rates of investment and development in different regions may create unrest and lead to the persistence of poverty unequally within countries. Increasingly, the distinction will not be between developed and developing countries, but between populations that are technologically adept and those that are not, and between those that are plugged in to rapidly changing knowledge and those that are not. But these changes are not elective; they already are becoming reality, and it will be the responsibility and in the interest of all to ensure that all countries and populations have access to knowledge and opportunity.

CHAPTER 3

Opportunities and Strategies by Sector

Over the next four decades, a doubling of the population coupled with economic growth will more than double the demand for food. Economic investment will have to rise by a large factor in order to provide plants, equipment, and jobs. The prices of some nonrenewable resources, fuels, and materials that are not economically recyclable will increase, denying their availability to many developing countries. The world's energy demands also will rise dramatically; in fact, in the next few years the expanding populations of India and China alone will account for two-thirds of the increase in total world energy consumption. In response to the growth in population, governments will be hard-pressed to provide the education and basic health services needed by the larger numbers of people and to maintain social safety nets for the growing numbers of the poor. Finally, all this economic activity will exact a toll on the quality of air, water, soil, forests, and other natural resources.

Yet the technological changes described here will present new opportunities for economic progress. Telecommunications, biotechnology, and materials science and technology will create new industries and new products—and thus new jobs. New fuels and new technologies for energy conversion will help to satisfy some of the demand. And other technologies will be applied to protect the environment. In fact, each of the major productive and social sectors of the economy

This chapter draws substantively on the invited papers by Balzhiser (energy technology), Baruch (technological innovation and services), Harwood (agricultural development), Heath (electronic manufacturing technologies), Lesgold (educational technology), Shine (health technology), and White (sustainable development), as well as the discussions of the break-out groups.

35

will be affected in different ways. In every sector there will be new opportunities as well as new problems—and in the mix a pathway for survival.

AGRICULTURE

Agriculture plays many roles in developing countries. It supplies food for rural and urban populations, generates export earnings, provides employment and forestalls migration to overburdened cities, and creates a critical interface between human activities and the natural environment.

No sector is so dependent on innovation and new technology to meet its goals as agriculture. Agricultural research is an integral part of the production system, and farmers in the developed countries are accustomed to relying on the research system and its representatives in the form of extension agents or the agents of commercial suppliers to aid in their constant struggle against weeds, pests, disease, and drought. Unlike, say, education, or even manufacturing, modern agriculture without innovation is difficult to imagine. To illustrate, for many years telephones were manufactured without significant modifications (and with no decline in volume or quality of communications), but without new chemicals or a new approach to pest management, crops would be destroyed by pests constantly evolving resistance to the old ones. Projections of future demand and supply of agricultural products depend critically on three important factors: population growth, income growth, and technology innovation. The first two are difficult enough to estimate on the basis of the factors involved, but the last can be extrapolated only from past rates of technical change. Ignoring it, however, creates a falsely (up to now) pessimistic picture of future needs.

As the demand for food continues to rise, together with the demand for other goods and services that depend on the same soil and water base, farmers worldwide, confronted with soil loss from erosion, will face powerful challenges requiring breakthrough technologies, some of which are not yet on the drawing board. Agricultural management strategies on the national and regional levels

---

Because the world is facing a shrinking land base and growing demand for agricultural products, the output per unit area of the food and feedgrains, as well as starchy vegetables, must more than double over the next 25 years. While there is considerable scope for increasing yields within the existing genetic potential, scientific breakthroughs will be needed to fully achieve the needed yields.

—RICHARD R. HARWOOD

also will have to be strengthened and revised, and many existing farms will have to be transformed.

High-resource, productive soils occupy only a small part, perhaps 10 percent, of total land area, but on these lands are the farms that can benefit most from the introduction of new technologies to increase food production. These farms can be divided into two categories: industrial-style plantation farms and multicrop integrated farms, often combining livestock and field crops. Plantation farms are usually dedicated to producing commercial crops, often for export. For that reason, they are generally better informed about and able to adapt new technologies, but they account for a very small share of food crop production in developing countries. Integrated farms, which are in better ecological balance, accounted for most of the production increases associated with the green revolution. They adopted the new technology rapidly once it became available in the agro-ecological region in which they were located. The most successful of them are knowledge-intensive and use the most advanced technologies, plant varieties, and management approaches, with on-farm sources of nutrients and pest control replacing many of the purchased fertilizers and pesticides.

Much of the land area of developing countries is marginal, characterized by low soil quality, low water availability, steep slopes, or high winds. This land area either belongs to traditional farms or is under forest. Although these marginal lands cannot support the continual cultivation of staple crops, millions of poor people depend on these lands for subsistence. Indeed, they carry the bulk of the rural population and must be kept viable until urban economic development can accommodate the excess labor force from these rural areas. Poverty in both urban and rural areas will place additional pressures on the land through the overexploitation and degradation of the fragile soil resource base.

The nature of exponential population growth is such that the greater increase occurs later in any time period. During the next decade, then, the world's food supply is expected to be stable, with only modest price increases. After that, however, new technologies and new resources will be required to sustain growth. But it will not be easy; the amount of land most able to support cultivated crops will most likely be decreasing because of soil erosion and such competitive uses as industry, housing, and roads. Increased food production to feed the growing population must come from yields. A new rice variety with up to 30 percent increase in yield was recently announced at the International Rice Research Institute in the Philippines. Yet much more research and development will be required to bring this technology to farmers' fields, and the inputs it requires must be provided without damaging the environment. Furthermore, if this can be done, then it must be done twice again in order to meet the demands of the mid-twenty-first century. And it must be repeated for other staples, starchy vegetables, and feedgrains. Since the plantation sector contributes little to developing country food supplies, and in some places is suffering the inefficiencies common to large vertically integrated enterprises in the manufacturing sector, new technologies must be developed with the needs of the integrated farms in mind.

It appears unlikely that much of the increased food required will be supplied by the oceans. Their present contribution to food supplies is less than 1 percent, and natural fisheries are in decline because of the misuse of technology and overfishing, as well as pollution from agricultural runoffs and other sources. It is hoped, however, that the expansion of aquaculture, aided by advances in biotechnology, might fill part of the shortfall.

Despite all the challenges facing agriculture, global trade patterns and communications technologies are creating opportunities for agroindustries in new geographic areas, including industries to produce industrial feedstocks from agricultural products. At the same time, science is developing substitutes for traditional agricultural crops. For example, cacao plantations may become uneconomic because of a new, lower-priced, "artificial" substitute for cacao, but some of them may turn to producing palm oil as a feedstock for a kind of biodegradable plastic. For developing country farmers, it will be a matter of analyzing and experimenting with the opportunities to find those that fit their specific soils, climates, markets, and capabilities. Education, investment, and the global exchange of information will be vital.

MANUFACTURING AND SERVICES

Manufacturing and services are among the sectors that have been most highly affected by radical innovations in computers and telecommunications.

Manufacturing

Among manufacturing industries, the electronics industry will probably be most transformed by advances in computers and telecommunications. It is, in fact, responsible for much of the new technology, and over the decades 1980-2000, this industry will have grown by more than a factor of six. Electronics manufacturing keeps pace with changes in silicon and integrated circuit technologies, which appear in the market as components or subassemblies. Most assembly plants use similar components, connectors, packaging, and power supplies—components that are seen at trade shows and traded openly to all manufacturers. Proprietary manufacturing technologies are less important, but knowledge of the technologies and of the market opportunities is crucial. They determine what is manufactured, as well as where, how, and by whom. Competition resides in design, including software, manufacturing capability, speed, and quality control—the first to market with a new product reaps much greater profits than the runners-up. This kind of competition is sometimes characterized as continuous creativity, continuous innovation, continuous productivity, or continuous learning. Because the lifetimes of electronic products are so short, the cost of materials is constant over the period and constant among different manufacturing firms. Thus the difference between a successful firm and its competitor is usually observed in the product design, product quality, manufacturing efficiency, and agility.

Modern electronics manufacturing is typically an integrated operation, coordinated with computer technology (also known as computer-integrated manufacturing, CIM). Manufacturing establishments depend on suppliers and assemblers for a large part of the value added, up to 75-90 percent of the value of electronic products. Suppliers generally ship small lots at high frequency so that they arrive just in time for use in order to save inventory costs and time. Such an arrangement requires that suppliers become almost coproducers, with databases and schedules linked through long-term relationships with the manufacturers. Linkages among manufacturers and suppliers are independent of location—that is, manufacturing plants in any country may depend on components and materials from suppliers in other countries, often on other continents. Integration is enacted through joint ventures, licensing, or foreign direct investment, and often involves technology transfer or other forms of alliances to share or jointly develop technology, even among firms in the same country. In the electronics industry, speed and demonstrated adherence to quality standards are vital, and computer-aided manufacture (CAM), high-performance teams of skilled workers, and automation may be more important competitive qualities than low-wage labor.

Since electronic product cycles are measured in months, not years, there is a premium on getting the initial designs right. In the earlier manufacturing style, design, manufacturing, and marketing were separate divisions of a company. Now it is recognized that these three activities must be coordinated and that marketing and manufacturing specialists must participate in the design process. The computer-aided design (CAD) process encompasses the ability to design-for-manufacture (to facilitate component fabrication and assembly) or design-

---

In the developing countries, policies and regulations on trade policies, soft financing, economic offset, and duties and taxes on the flow of manufacturing elements can significantly affect the climate for manufacturing. Furthermore, the competitive complexion of a manufacturing entity is becoming increasingly dependent on its successful utilization and leverage of its global production network. Without the support of the developing countries themselves, the manufacturing operations will not be globally competitive.

—SIDNEY F. HEATH III

for-environment (to reduce unrecyclable wastes or pollution in the manufacturing process). When the right factors are present, the most efficient solution might well be characterized as "design-for-developing country participation." It might be based on proximity to markets, access to raw materials, low wage rates, or new applications of products in developing country markets.

Services

Services—usually defined as intangible and nonstorable goods, which are generally insulated from international competition but highly prone to regulation internally—are being transformed by the changing nature of competition in service markets. Some new categories of services—transborder services—are explicitly international; others are embedded in the value chains of international integrated manufacturing. Because the distinction between manufacturing and services is blurred by these categories of services with their strong international components, local regulation of services will likely become less effective.

The growth in transborder services has stemmed from cheap computers, low wages, and low-cost, efficient telecommunications. The cost of electronic com-

---

Among the so-called services, a country's ability to manage and use information will be the single determinant of its rate of development.

—JORDAN J. BARUCH

---

munication is becoming independent of distance, and the prices of memory and bandwidth are falling rapidly. Data-entry, translation, and financial services are examples of the services that have been exported from industrialized countries. For example, when an 800 number is dialed in the United States, the voice that responds often will come from the Caribbean.

It is likely, however, that many of the present transborder services being performed in developing countries will decline in importance or disappear. For example, as voice recognition technologies become more effective, data-entry personnel or telephone operators will be replaced by a smaller number of "native speakers," and these types of services may move back to the developed countries or to those countries predominantly English-speaking. At the same time, the number of direct-order businesses for export may increase, favoring the developing countries—another blurring of the distinction between manufacturing and services. Some services such as translation, software development, insurance claims analysis—those mostly requiring a skilled work force—will not decline. Services that support foreign direct investment, or information networks, or integrated manufacturing will provide fresh opportunities.

Embodied services are another class of service that is frequently performed in a developing country for clients in developed countries. Some embodied services are part of the integrated manufacturing process but differ from manufacturing because the service firm performs none of the functions normally associated with manufacturing such as design, marketing, or engineering. Research and development and custom manufacture on demand are examples. Particularly in the garment industry, manufacturing firms outsource the actual fabrication of goods while retaining the design, engineering, and marketing functions. It is hard to distinguish custom fabrication of specified garments on order from the more traditional services such as cleaning and packing or transport.

THE ENVIRONMENT AND ENERGY

The earth's environment will never again be able to sustain the energy use, pollution, and resource waste that characterized the development of the industrialized countries to their present levels of consumption. Today, most of the world's population increase, the highest rates of economic growth, and the resulting threats to the environment are occurring in the developing countries. But that is also where most of the humid tropical forests that help to sustain the balance of atmospheric carbon dioxide are located, and where the congested and polluted cities and the falling water tables have as yet found no remedy.

In the past, environmental problems were considered the inevitable side effects of development. Some technologies were "cleaner" than others, but increased productivity was the primary goal, and any environmental damage was either taken care of separately or disregarded. Some of the excesses that occurred in the Soviet Union and the Middle East were consequences of that philosophy. Fortunately, the public's attitude has now changed; environmental science is a new discipline, and environmental technology is a $300 billion a year business worldwide. Manufacturers now "design for environment," using informatics technologies, computer-aided design, and computer control of manufacturing. The results are processes that pollute less to make products that are easier to recycle.

The present great wave of new technologies and technological concepts collectively represents a new environmental technological offensive. Properly directed and financed, this offensive could open pathways to an environmentally sustainable future as well as restore damaged environments. Technological innovation by itself is a necessary, but insufficient, means to that end.

—ROBERT M. WHITE

One key to sustainable development—development that does not cause permanent degradation of the environment—is the kind of energy source used for domestic and industrial applications. Energy sources can be divided into two categories: commercial sources and traditional or informal sources. Commercial energy includes electricity, refined engine and heating fuels, batteries, and other manufactured devices for producing energy. Traditional sources are firewood and charcoal, dung, water wheels, and animal traction. Half the world's people have little or no access to commercial energy. Modern commercial energy technology is capital-intensive, and most of it is driven by fossil fuels. The reserve-to-production ratios for the world's commercially proven reserves of oil and gas are currently around 45 and 65 years, respectively, while those for coal (from which gas and synthetic liquid fuels can be made) are over 200 years. Given these estimates, it is likely that abundant oil, coal, and gas will dominate the commercial energy picture for decades, barring some unexpected new source of energy.

Today, the efficiency of power plants as well as that of electrical devices is climbing. The best gas-powered combined-cycle combustion turbines approach 50 percent higher efficiency than direct fossil fuel-fired power plants. Other technologies are based on coal, which constitutes 80 percent of the world's fossil fuel resource and is the main resource of the two largest countries, India and China. Integrated gasification combined-cycle (IGCC) technology first gasifies the coal, then removes the impurities in the gas, and sends the gas to a combined-cycle turbine. These turbines, manufactured worldwide, have relatively low capital and fuel costs, low emissions, and a modular design. They are fabricated in a variety of sizes and shipped by railway. Other coal-based technologies, such as fluidized-bed power generators, also are relatively clean. For fossil fuel-powered electrical generation, the fuel is abundant and efficient technologies are available. The major constraint relates to the environment and to special situations in which access to fuel is difficult or expensive.

The availability of modular units is changing the nature of power generation and distribution. Modular technologies are generally small, factory made (possibly including local components), and quickly installed. Moreover, they can be installed close to the load to minimize transmission and distribution costs. Similarly, new macro-electronic technologies for control over long power lines are under development, and small, dispersed generating units are an increasingly viable option. These technologies already are having an impact in the United States, where many utilities are redefining themselves as transmission and distribution companies. In fact, the energy industry appears to be heading in the same direction once taken by the computer industry, which passed from mainframes to desktops to laptops to networks of small units, which ultimately have the power to challenge the largest supercomputers.

The main problem related to energy generation at present is its impact on the environment. The threat to soils and forests is a function of the continued use of the traditional biofuels such as firewood, charcoal, or dung, where electricity is unavailable or costly. Making electricity available to all populations is a major element of the long-term solution to deforestation and erosion. Another major environmental concern is global warming. Because of the difficulties and the costs associated with nuclear power for electricity generation and the inevitable production of carbon dioxide in fossil fuel combustion, the main long-term option for stabilizing greenhouse gas emissions is renewable energy sources in the forms of solar power, wind power, and biomass energy. Presently, these sources are more expensive than fossil fuels, but their costs are dropping. In 15 years, the costs per kilowatt-hour for renewable technologies have fallen between 30 and 200 percent, depending on the technology. Although renewables may soon be commercially competitive everywhere, for the moment they are a realistic alternative only in remote areas or where fuel costs are very high.

Photovoltaic technologies have the greatest potential in the tropical regions. The units, which are modular, can be easily transported, assembled, and maintained in isolated areas, as successfully demonstrated in Brazil, Indonesia, and other countries. Photovoltaics remains a promising area for research and development, part of the silicon revolution of rapid innovation. The present price is $0.30 per kWh, compared with the competitive price of $0.05 per kWh, but a new 20-kW system under construction in Arizona may approach $0.10 per kWh.

Solar thermal generators use large collectors to concentrate the direct heat of the sun for conversion to electricity. Their size and capacity put them into competition with fossil fuel combustion turbines, which are cheaper and more efficient. This technology might be competitive in a tropical region without coal or gas.

Wind turbines, presently competitive in regions with persistent winds, are being used in the United States and parts of Europe. The variable-speed turbine is capable of producing electricity for about $0.05 per kWh, and the fierce competition that currently characterizes the industry may result in even lower costs in the next decade.

As the next century unfolds, the issue of global sustainability will begin to transcend the separate concerns of population, energy, economy, health, social welfare, and the environment. New means of achieving sustainable development will be required, and efficiency is likely to act as the backbone of all future strategies of sustainability. In this, electricity will be important in reconciling human aspirations with resource realities.

—Richard E. Balzhiser

Solar and wind technologies produce energy when the weather is favorable, but much of the consumption takes place at other hours. Thus once energy storage is improved, renewables should be better able to compete effectively with advanced turbine technologies. Promising technologies for energy storage include fuel cells whose feedstock is produced with renewable energy, thermal storage devices, and advanced batteries. These devices are presently relatively expensive and inefficient, but they are a fertile area for research and development.

Finally, sources of biomass for commercial energy generation are either waste products or crops grown specifically for energy production. In either case, this area may benefit from advances in biotechnology, both for higher productivity of fuel crops and for more efficient conversion of the biomass to energy. The material can be either burned directly or converted to a biofuel such as ethanol or methanol, liquid or gas, for use in advanced engines and combustion turbines, although under present conditions this is not economically attractive.

HEALTH

The markets do not work well for health. In the health systems of most countries, the consumer—the patient—has very little knowledge, and therefore little basis for choice, of the type and quality of care received. Rarely are there alternate providers, and the costs of care are borne by a third party, either the government or an insurance company. For similar reasons, investments in health-related research and development do not respond directly to the needs of the patients in most countries or of most providers. Research and development are largely carried out by big international firms—American, European, and Japanese pharmaceutical or equipment manufacturers—that make half their profits in the United States. Moreover, the United States has a commanding lead in fundamental research with an annual investment of $25 billion. It is not surprising, then, that most research efforts are aimed at the U.S. market. The situation is sharpened by the fact that the United States, mainly through its Food and Drug Administration, has a rigid and costly process for the approval of new technologies—but one that emphasizes efficacy and safety and ignores cost-effectiveness—and a legal system and tradition that awards high payments for liability claims against companies that produce health technologies.

The result is that research and development in the health industry do not generate the technologies that serve the needs of the developing countries, mostly because those needs do not presently involve lucrative markets, although the number of affected people is large. Some developing country needs do coincide with areas of intense research and development efforts in developed countries, but the fruits of that research are not reaching Third World nations, possibly because they involve expensive, high-technology solutions that are beyond the means of poor populations. Some examples of these research and development needs are:

- *Reproductive health technologies.* Developing countries especially need contraceptive technologies for men and women, as well as methods that simultaneously protect women from infection since they bear the heaviest burden from sexually transmitted diseases.
- *Micronutrient supplementation.* Cost-effective delivery systems are needed for providing vitamin A and other essential diet supplements to children, whose health is the most vulnerable to nutritional deficiencies.
- *Vaccines.* Immunizations will protect children against the most common childhood diseases and against diseases that carry large burdens of morbidity or mortality in the tropics.
- *Primary health care.* Expanded facilities are needed for primary and outpatient clinical care, including the use of cost-effective diagnostic and treatment technologies through telemedicine.
- *Chronic diseases.* Cost-effective interventions are required to control the growing prevalence of chronic illness—heart disease, cancer, stroke, lung disease, and diabetes—and to reduce the use of tobacco, which exacerbates these illnesses.
- *Information and surveillance technologies.* Such systems could anticipate the emergence and spread of little-known or local diseases and drug resistance in common diseases.
- *HIV.* Research could tackle the problem of behavior modification to prevent the spread of the disease and monitor new strains of HIV and different transmission patterns to complement the large research programs of the advanced countries.

Another area in which current research is not addressing major problems in developing countries, or developed countries for that matter, is behavioral research. Many of the most serious threats to wellness in all countries involve behavior: addictive and hallucinatory drugs, mental disorders, and most especially violence. Neuropsychiatric illnesses, together with cardiovascular diseases and injuries, deliberate and accidental, constitute half the disease burden of the developing countries.

An instructive example of the bottom-line bias of today's health-related research and development is vaccine development. Vaccines are the most cost-effective technologies known, yet for technology companies they generate smaller profits and higher potential liabilities than treatment technologies that are used over a long period. This is illustrated by the decision of a consortium of U.S. pharmaceutical companies to make a united effort to develop antiviral agents against HIV infection rather than a vaccine to protect against AIDS. Similarly, research on contraceptive vaccines or devices badly needed in developing countries is impeded by political opposition in the United States. In short, health care—and health care costs—in the United States are dominated by expensive high-technology diagnostics and treatments. Thus these technologies and this system are not necessarily the most appropriate for transfer to the developing countries.

This being said, the needs of the developing and developed countries have coincided for two advances that provide services to large populations at reduced costs and that have their origins in research areas touched by the computer and telecommunications revolutions. The first advance is improved outpatient services. Many procedures once requiring prolonged hospital stays are being performed in outpatient clinics with access to information resources, and average lengths of hospital stay, even in tertiary care facilities in developed countries, are dropping rapidly, driving down costs and decreasing risks of nosocomial infections. Today, in fact, many hospitals are able to eliminate beds and replace them with ambulatory facilities and satellite clinics, supported by nearby lodging for patients and their families.

The second advance is telemedicine, the name given to systems in which a central facility staffed by physicians is able to diagnose and treat patients in remote locations by means of computers and telecommunications technologies. Patients visiting the remote locations are "examined" by physicians through interactive video and communications equipment that allows them to view patients, receive diagnostic data and x-rays, and instruct attending physicians or paramedics in treatment procedures. In the future, virtual-reality technologies, combined with fiber optics and endoscopy, may enable physicians, working in a central facility and provided with global information resources, to perform surgery and other operations in remote locations. Even in the near term, in both developing and developed countries, the most appropriate health care model will be ambulatory diagnostic and procedures rooms with a small number of hospital beds supported by high technology and remote telemedicine services.

Investments in health research pay large dividends, but presently only 2 percent of health research is carried out in developing countries, despite the fact that 93 percent of preventable mortality is found in the Third World. Only one international health research center is located in a developing country, the International Diarrheal Disease Research Center in Bangladesh. Its operating cost is approximately $10 million per year, a small fraction of what the cost would be if it were situated in a developed country.

EDUCATION

In developing countries, the per capita expenditure on education is about half that of member countries of the Organization for Economic Cooperation and Development (OECD). School standards are low, and in many countries the goals of achieving 100 percent primary education and female literacy remain unfulfilled. Modern technology, which is not usually a prominent feature of the classroom in any country, is scarcely used. Educators, however, often cite poorly prepared, overworked teachers and poorly equipped, crowded classes as the rea-

Because they often have more educational chores to accomplish and less money to invest, the developing countries especially could benefit from the technological leverage of learning. But because educational systems also are extremely stable and resistant to change, it is important to establish clearly whether a given technological contribution will sufficiently enhance educational productivity before undertaking any major effort to use it.

—ALAN M. LESGOLD

sons why new technologies would not work and why therefore they should not be introduced. Some developing countries have noted as well that there are large numbers of computers and other technologies in U.S. schools, but that their influence on productivity has yet to be demonstrated.

Education is highly sensitive to local culture and language. Educators know that they must build on what students already know and how they think, and that in most cases the students must be taught in their own languages. Countries can borrow tools and technologies from outside, but they must redesign them, integrating them with the local culture and knowledge base if the new techniques are to be accepted and sustained. It has been demonstrated, however, that some advanced technologies, interpreted in the broadest terms, can have important, cost-effective yields even though they may seem too exotic for developing countries. Hardware development—textbooks and computers—is far ahead of software, but such soft technologies as theories of learning, educational standards, and translation and voice recognition programs can make important contributions.

Schools are generally resistant to technology-driven changes. Teachers are protective of their status in the classroom and their position within the system, and they must be fully comfortable that a technology will enhance their role, not supplant them, before they are likely to employ it willingly. Because they also are often overworked, teachers must be confident that a technology will work and will not cost them extra effort. Teacher training and equipment maintenance are therefore key requirements for the adoption of any hardware-based technological solution.

As for the benefits of such solutions, electronic networking can provide teachers with practice and technological experience and put them in contact with other teachers. It also can provide materials they can use in the classroom, either prepared centrally on a national level, or, perhaps through the Internet, from international sources. Use of World Wide Web would be particularly valuable; it has found favor with teachers in developed countries and is a popular resource for students.

The technologies appropriate for mass education—such as radios—must be readily accessible to all and like any new technology in the marketplace must be evaluated by the learning they produce. In general, low-maintenance technologies that are easy to teach and easy to use work best. One example is interactive radio. Programs are designed to reinforce lessons by encouraging pupils to respond aloud. Pupils listen in a group and respond together to the instructions or questions on the program. This approach is especially comfortable to many children who like the oral approach and like to work in groups. Radios are ubiquitous, and this technique has been shown to be effective and cheap, with costs similar to those for standard textbook-based alternatives. Another useful technology is the computer-based printing system, linked to a central network. Such a system allows print masters to be made locally and large quantities of texts and other materials to be printed cheaply. And not to be forgotten, CD-ROM, combined with a laptop computer and a wireless modem, presents many possibilities, including low-cost video production and access to material via World Wide Web. The costs of these items have fallen drastically, which may make them realistic options, even for rural and remote area schools.

Science education is a major part of producing a technologically literate citizenry. But competent science teaching requires knowledgeable teachers, the proper laboratory equipment, and a good syllabus. Technology can sometimes provide substitutes for all of these elements. For example, the National Science Resource Center of the National Research Council and Smithsonian Institution has prepared syllabus materials for the primary and secondary levels that provide units of experimental science and include low-cost materials and supporting information for teachers. For universities, the concept of the virtual classroom allows developing country universities to take advantage of materials prepared at the best universities in the developed countries through multimedia technology resources. Many aspects of laboratory instruction can be simulated by computers for students.

Most of the innovative technologies that are transforming the manufacturing and service sectors place a premium on the skilled, trained worker at the expense of the unskilled, low-wage worker who in the past has made many of the developing countries competitive in the marketplace. Countries that hope to remain competitive or to enter new niches must therefore put a high priority on workplace training. Fortunately, many of the new technologies may themselves be used to enhance worker training. For example, firms can utilize their own computers and computerized tools in the workplace to train potential workers during the after-operations hours. Coached apprenticeship, which provides training and assistance to the worker at his or her own workplace, is sometimes called just-in-time training because the knowledge is imported at precisely the time and place it is needed and is thus most effectively absorbed and assimilated. A similar technique, intelligent coaching, also is computer based. It uses audio recognition of known text to enforce correct responses and to correct errors. The combination of these self-paced techniques with the guidance of a mentor may be the most powerful tool of all.

The formal design and adoption of educational technologies, whether newly conceived or merely new for the country, will require research and development at the local level. Teachers should participate actively in these experiments, guided by experts, and share experiences with other teachers. Regional or national centers, in collaboration with ministries of education, could effectively guide the process of adapting technologies to local use by mobilizing teachers, providing them with technical assistance and modest sums for materials, and analyzing the results. Once the new technologies are adopted, training teachers in their use should be built into the curriculum.

Training already is an established field of research in which the private sector has played an active part. The field could expand to encompass even more innovation and technology transfer if firms had the incentives to invest in training activities. In many developing countries, though, fear of hijacking (recruitment of workers who have been trained at the expense of another firm) has discouraged training initiatives. Regulation (or deregulation) of the transferability of retirement and health benefits, as well as government-led initiatives to encourage cooperation among firms for training, would help to mitigate this problem.

This chapter has described how advances in technology have affected various sectors, and where needed technologies are lacking. Changes in world markets and the impact of new technologies are largely beyond the control of any authorities or governments, although some governments have influenced the pace of development and implementation of selected technologies through tax incentives, partnerships, and government-funded research and development. Generally, the private sector is leading the way in the technology revolution and controlling the application of new technologies. Thus in those sectors that are generally public—notably education and, in part, health—even some of the technologies available and clearly needed are not being applied. The final chapter is therefore devoted to some recommendations for action by governments, the private sector, the scientific community, and the development agencies, led by the World Bank, to ensure that a majority of the world's population benefits from the technology revolution.

CHAPTER 4

Rising to the Challenge: Priorities for the Developing Countries and the International Development Community

The changes that will result from the new technologies described at this symposium will present a different face to different countries. Many countries simply do not have the minimum levels of capital, infrastructure, human resource capability, basic services, and technological awareness to benefit from the telecommunications/computer revolution over the short term. Many of these countries, however, are the very ones that will have to rely most on technology to relieve the pressures on their food supply, health and education services, and environment that will accompany the next doubling of the world's population. Ironically, while the rapid changes fostered by today's sophisticated telecommunications and computer technologies are likely to become reality—and many of them already are—those still required for survival remain on the drawing board. The international development community must, therefore, not abandon its efforts to assist these least-developed countries. The telecommunications/computer revolution may, in fact, have little to offer them.

For other countries, however, there are opportunities to grasp that would allow them to catch up—with the developed countries, with the world economy, with their own environmental remediation requirements, and with the demands of their own growing populations. But to succeed, they too will require some assistance in planning, some awareness-raising, and substantial changes for investments in economic growth.

This chapter seeks to identify some of the priorities, delimited by problem area, that will permit developing countries to accelerate the process of applying

---

This chapter draws substantively on the invited papers by Baruch (technological innovation and services) and Shine (health technology), as well as the discussions of the break-out groups.

the new tools offered by the technology revolution. Most of these priorities must be implemented by the countries themselves, the private sector, and the scientific community. Others will be part of an international effort coordinated by development agencies and the World Bank. This chapter will conclude by examining the corresponding new roles for the cast of actors on the international stage.

FOOD SUPPLY

The population pressures that will affect a significant number of the world's countries in the early twenty-first century give priority to issues of food security. The increases in crop and livestock production that will be required, estimated at greater than 100 percent over the next 40 years, will have to come mainly from gains in yields per hectare, mostly on land already under cultivation. This will require technological advances in agriculture that can be compared in scale and impact only to the green revolution of the 1960s—a revolution that was largely led by international research centers located in developing countries, funded by international organizations, developed countries, nongovernmental organizations, and host governments. But since the sixties the international political climate has changed. The global markets are no longer dominated by so few countries, and the private sector plays a greater role in research and development. Nevertheless, the remedy for projected food shortages remains the same: more technology.

Basically, three farm types are found in the tropics: the industrial-scale or plantation sector; integrated, multicrop farms; and the traditional farms, usually found in marginal-soil or low-resource areas. In general, private market incentives should suffice to spur growth in the commercial plantation sector. Thus efforts to promote increases in the food supply should focus on productivity and concentrate on the most competent, integrated, multicrop farms found in high-resource areas. New technologies must be introduced on these integrated farms; newly strengthened national research and educational institutions could help to develop or transfer the technology and support productivity gains. Such measures also might have an impact on the traditional sector if efforts are made to study and improve traditional crops and farming systems. Assuming the new technologies will be forthcoming, the key problem will be getting them into the hands and the minds of the farmers—technology transfer.

Market incentives for the adoption of biotechnology-generated products generally reach the plantation sector. But the integrated farm sector has less access to information and may be more conservative about new technologies, having less margin for survival. Promotion, demonstration, and extension may be essential to the further adoption of new cultivars and improved varieties. Investments in soil productivity enhancement, associated with crop rotations, conservation tillage practices, increased efficiencies of fertilizers and pesticides, and improved farming practices to avoid erosion and runoff, will provide major gains. For the traditional sector, scientific breakthroughs that might permit the cultivation of high-value commodities on marginal lands without unduly degrading the environment would make a major impact. This problem, however, holds little interest for the developed countries, where most biotechnology research is carried out. Solutions, then, should be pursued in national and regional agricultural and biotechnology research centers.

PRODUCTIVITY AND COMPETITIVENESS

It is in manufacturing and services that developing country firms perhaps have the best chance of using advanced technologies to propel them into equal participation in global markets. But two conditions must be satisfied for these firms to become integral parts of global integrated manufacturing networks: the developing country firms must be competitive in price, timeliness, and quality; and the policy environments of their countries must be conducive to global production.

The notion of "design-for-developing countries" will work if these countries are able to offer access to large or emerging markets, or to sources of materials that cannot be readily obtained elsewhere. For example, a developing country firm could introduce a product—such as an electric water pump designed for a consumer appliance—to a new market for use in tandem with a solar generator. Another firm might provide workers with a moderate level of education and skills at relatively lower wages. Or a firm might offer boutique manufacturing because it is small enough and flexible enough to custom manufacture on demand and in small quantities. Consideration of the new patterns in collaborative manufacturing will help to identify many opportunities for manufacturing enterprise development in developing countries.

In much of the developing world, the industrial sector is dominated by a few very large firms and a large number of cottage industries. Middle-sized companies that have the flexibility and capability to serve as reliable suppliers in a value chain are in short supply. For such countries to compete, new government-developed incentive programs could encourage the formation of such mid-sized firms to serve as a supplier base, thereby reducing imports of subassemblies and components. Proximity of suppliers is a valuable asset, as is the timely import of high-technology components such as displays or integrated circuits that are not manufactured locally. Similarly, a good transport infrastructure, including rapid customs services, is important. The establishment of industrial parks and duty-free zones would encourage participation in global networks.

For some countries, substantial changes in the domestic economic environment—such as low inflation, deregulation, and intellectual property rights protection—and a commitment to change by both the government and the private sector will be needed to create the climate for effective participation in global networks and markets. But any such changes should be undertaken as part of a vision of the future created and articulated by each country. All sectors—government, private sector, universities, labor—should contribute to that effort and then identify and carry out the actions to achieve that vision. A mechanism useful for this purpose is a national government-university-industry roundtable at which government officials, business leaders, scientists and engineers, and economists exchange ideas and prepare a plan for public debate and government decision. Nongovernmental organizations, particularly trade associations and professional associations, also should play a part. The knowledge required for rational decision making in this area is not exclusively held by governments. The challenge presented by the technology revolution requires participation by all sectors of society.

Such a vision should be supported by a strategy to fulfil it. A country may decide to enter global manufacturing markets in areas where it has a comparative advantage by first offering peripheral components or services such as software, spare parts, field services, tools, or postharvest processing of agricultural products. These arrangements could be consolidated by license, joint ventures, foreign direct investments, or government-required offsets on other contracts. Such ventures will spin off knowledge of the technologies and of the international markets. Later, the country may wish to introduce products to local markets, enter into long-term partnerships with major international producers, or undertake complete product manufacture in competition with producers in other countries. Each country must be firmly aware of its own interests and recognize that no one gives away anything valuable. Today's technologies may be available for license, but only when tomorrow's are on the test bed.

Governments, in collaboration with the private sector, can take several kinds of actions to implement their visions. For example, in a competitive environment low-cost labor will by itself count for less, but a moderately low-wage yet educated and trainable work force may have a definite competitive advantage. Even so, the cost to a firm of training may offset low direct labor costs, and thus it may be advantageous for government to share training costs and to take measures to discourage the practice of one firm hijacking workers trained by other firms. Government also could create institutions that offer training in the management of technology, possibly following the model of the institute set up through joint U.S.-Chinese cooperation at Dalian. Other services that would assist local firms are: technology scanning and forecasting, technology demonstration and transfer institutes, trade shows, and technical information services linked to worldwide information networks. Information also might be distributed through the facilities of other public services, such as electricity and telephone, that reach most establishments.

Deregulation and the elimination of trade barriers are important steps along the path to competitiveness. New informatics technologies are, in any case, eroding the capability of regulatory agencies to control the service industries, or to prevent domestic services from being marginalized by international competitors. An example is modular telephones, which can undercut local telephone services and provide untaxed and unregulated international service as well. (In Chapter 2, however, this example is used positively—as a way developing countries can jump-start a telecommunications industry.) Obstacles to foreign direct investment and to high-technology imports can impede local participation in global markets. A closer look at the experience of the newly industrializing countries of Asia and Latin America would be useful. The World Bank and other neutral parties are ready to advise developing countries in the area of regulatory reform.

The success of many of the actions described here will depend on cooperation from the developed countries, from investors who see a profit in working in the developing countries, from manufacturers seeking partners, and from scientists and engineers who can provide valuable advice. Communication among all these players is essential, and this too could be facilitated by modern informatics technology. On an Internet bulletin board, questions can be posted and answers returned, and third parties can comment on both the questions and the answers. A bulletin board devoted to the technological aspects of development could serve as a forum for questions from developing countries on technological opportunities and answers from investors, potential partners, and scientists and engineers, with commentary from the World Bank and other international monitors. This initiative would cost virtually nothing and could be either informal and open to the public or restricted to a defined group.

ENVIRONMENTAL AND ENERGY TECHNOLOGIES

Technology will contribute in different ways in the environmental arena. Pollution and resource depletion, the two faces of environmental destruction, may find some sources of remediation among the emerging technologies. New techniques for bioremediation can clean up some of the damage to fragile environments, such as oil spills or heavy metal contamination. Similarly, other technologies, developed in the West in response to public agitation over hazardous waste sites and industrial pollution, can reduce the emissions and clean the effluent from industrial plants. These technologies can be licensed and applied in developing countries before extensive damage has been done.

Energy generation and use are a major source of pollution, but many new and emerging technologies for efficient, cleaner electricity generation are available and well suited to developing countries. In fact, this is a good time for developing countries to add new, more efficient low-emission technologies to their capital stock while per capita demands are still low. Based on current demand projections, in 20 years the developing countries will require a tenfold increase in generating capacity as well as end-use equipment. Fortunately, these countries are in a position to leapfrog to a new generation of technologies that is far more efficient, less costly, and less polluting than was historically available to the industrial countries, even at a far more advanced state of development. The result will be an approach from below to the present-day optimal level of energy use per capita for the given size of an economy, instead of the path of expensive retrofitting required in many developed countries.

During the oil crisis of the seventies, many international organizations and bilateral donor agencies undertook extensive energy planning, assessments, and research programs. Today's situation may be just as critical in the long run. Many countries, finding themselves undercapitalized in the energy sector and with growing demand, are making key decisions that will affect the economy and the environment for a long time. The issue this time is not so much the choice of fuel—although there are still choices to be made among fossil fuels, renewables, and nuclear—than that of generating and end-use technologies and their relative benefits in terms of high efficiency, decentralization, and pollution control. The scientific community, supported by donor and development organizations, should take the lead in providing good offices and advice for these important decisions.

The great importance of macroeconomic stability, of pricing and tax policies based on economic principles, and of a satisfactory regulatory framework for investment, is well known. Government must play a key role in sending consumers the right signals and enabling markets to function efficiently. This role should embrace such approaches as (1) setting standards and codes (for example, for the performance of buildings, appliances, and equipment); (2) monitoring pollution, establishing environmental standards, and introducing environmental taxes, laws, and regulations (such as for phasing lead out of gasoline); (3) attending to property rights issues, which can be important for the design of environmental policy (rights of the polluting and the polluted parties) as well as for investment; (4) providing investment incentives for the adoption of new and innovative technologies; and (5) providing for the sharing of risks, especially important in countries with undeveloped capital and insurance markets.

RESEARCH AND DEVELOPMENT

The research and development community has brought the world the computer/telecommunications, materials, and biotechnology revolutions. A creative partnership of the public and private sectors in the Western countries and Japan produced most of the new technologies discussed in this proceedings and is in the process of changing the world. One would not assume that the research and development system needs any remediation, but in many sectors it is not, in fact, providing the innovations and discoveries required for the problems of most importance to developing countries. Most fundamental research is done in and by the developed countries, and much of the applied and developmental research is carried out with their large and affluent markets in mind. The needs of the poorest countries or of some tropical countries—especially in agriculture, health, education, and the environment—often are not considered. This is in a sense a "market failure" in research and development, and some "intervention" may be necessary.

The most successful model for research and development for the benefit of the developing countries is the system of research laboratories of the Consultative Group for International Agricultural Research (CGIAR). These laboratories, most located in developing countries, specialize in corn and wheat (Mexico), rice (Philippines and Liberia), forestry (Kenya and Indonesia), livestock (Ethiopia and Kenya), insect physiology and ecology (Kenya), tropical grains and legumes (Colombia and Nigeria), arid-zone agriculture (India and Syria), potatoes (Peru), training (Netherlands), food policy (United States), and other areas of agricultural research. Their successes stem from their concentration of scientific resources, including the skills of researchers from all countries, on local problems in the developing countries. Some of their weaknesses have been related to difficulties in disseminating and communicating their findings. The lessons revealed by several assessments of the CGIAR research network should be studied with care before the model is copied or expanded.

Other areas that could benefit from regional centers of excellence in research and development are: health research, with a concentration on vaccine development, contraceptive technologies, and tropical diseases, as well as the capability to track emerging diseases and drug resistance; energy research, with an emphasis on adapting renewable source technologies to local conditions; environmental research, to understand and minimize the impact of agriculture on the tropical environment; and education, with a focus on technologies for mass education in poor countries. The new centers of excellence could be built on existing research centers—either national research institutes, which exist in many countries, or the CGIAR research centers. They would offer professional training to developing country researchers and an environment that would attract distinguished researchers from both developing and developed countries for limited stays. Research in the social and behavioral sciences should be included in the programs of these centers.

The proposed centers might differ from the centers of the CGIAR system, established in the 1960s, by utilizing telecommunications and computer technologies to involve the private sector and a wide spectrum of scientists working in their own laboratories. This different kind of international research network, appropriate to the 1990s and into the next century, should be explored thoroughly by the scientific and donor communities.

Many developing countries have been sending scientists, engineers, and other professionals overseas for training for decades; China now has more Ph.D.s than England, France, Germany, and Japan. Some say, however, that these developing country professionals are not yet producing up to expectations. A similar situation occurred before World War II when the United States had more scientists than Europe, but they too were not making a great impact. The war focused the efforts of the Americans, and a radical but more benign upheaval might serve to energize the scientific elites of the developing world in a similar way. The stimulus could be provided by modern information-computer-telecommunications technology, coordinated by regional centers of excellence to give these scientists access to up-to-date technical information and publications and allow them to communicate easily with their colleagues around the world. These scientists might, then, provide the leadership needed to bring their countries into the technological mainstream.

INSTITUTIONAL ROLES

The developments suggested here will not come to pass unless all sectors of the world community do their part. The challenge of marshaling technology for development will require new roles for governments, the private sector, research and scientific institutions, and the World Bank and the development community at large. These new challenges are described in the rest of this chapter.

Governments

- Maintain awareness of the profound influence that technological changes may have on the global economy. Initiate a planning process involving all social sectors to create a vision of the country's role in a new global market and take steps to implement that vision.
- Create a legal and economic policy framework that encourages innovation and provides firms and individuals with the ability to respond to technical change in an agile way. New technical and information-oriented institutions and technical assistance programs, especially related to quality management, may be vital. Provide incentives to the productive sector to respond to opportunities for small, technically-oriented companies.
- Invest in the physical and technological infrastructure, especially communications and transport, needed to enable the productive sector to acquire and put to use the most appropriate and effective technologies, seeking private sector participation where possible.
- Bring technology to bear in the provision of public services, in particular to reduce the cost and increase the quality and coverage of educational and health services. Consider investments in new technologies for energy generation that are more efficient and less polluting and in technologies for cleaning up the environment.

Private Sector

- Maintain awareness of technological advances in industry and acquire the most effective production methods and products through research and development, international agreements, joint ventures, and imported technology. Gain the capability to access knowledge through international networks. Be aware of quality management requirements in international markets and reorganize procedures and facilities to achieve quality standards.

- Adopt organizational changes to better manage intellectual assets, invest in innovative activity, and improve quality control. Because most technical change comes from incremental innovations on the factory floor, be open to employee-initiated changes.
- Recognize the importance of employee training to incorporating new technologies and converting knowledge to value. Be prepared to join forces with the government and other private firms to leverage resources for employee training.

Scientific and Research Community

- Take a leading role in advising developing country governments and the development community of new technologies and their implications for developing countries. Participate in information clearinghouses on the Internet to assist researchers and producers in developing countries.
- Identify research priorities for regional and national research centers, addressing the needs of developing countries. Assist and encourage research and development on the local level to encourage the application and adaptation of new technologies in specific developing country contexts.
- Form partnerships with research institutions in developing countries and encourage research partnerships across developing countries.

Development Community

- Put technology issues at the forefront of individual country development assistance strategies. Help the least-developed countries adapt to the changes brought on by the new telecommunications and computer technologies.
- Raise the awareness of developing country governments and other donors of the opportunities and challenges offered by new technologies—for example, by supporting seminars and studies on the implications of technological change for developing countries.
- Play a connector role, forging partnerships between developing countries and the scientific and research community to increase access to knowledge and apply it to developing country problems.
- Help to make information on technologies more widely and easily available to developing countries. Explore options for providing information facilities—for example, on energy and environmental technologies—via networks such as Internet.
- Provide honest broker services, an advisory role that could be performed in conjunction with national scientific academies or other scientific organizations, to assist governments to evaluate different technologies.
- Finance pilot or demonstration projects that apply new technologies in specific developing country circumstances.
- Assist developing countries in managing the negative impacts of change brought by the technology revolution.

INVITED PAPERS

The Global Generation, Transmission, and Diffusion of Knowledge: How Can the Developing Countries Benefit?

GEORGE BUGLIARELLO
Chancellor, Polytechnic University

It is by now almost a mantra that the human species has reached a stage in its trajectory in which information, or more broadly knowledge, has become the new *leitmotif* and organizing principle of society, much in the same way that energy was for the industrial revolution. (When feasible, the terms *knowledge* and *information* are used interchangeably in this paper, although, strictly speaking, knowledge connotes more than quantifiable information; it involves awareness, insight, and the power of discernment.)

Information and learning—the process by which information is received and absorbed to become knowledge—have guided human actions from the beginning. But it was only until well into this century that engineers and scientists were able to define information as the removal of uncertainty, to measure it precisely, and to create devices to transmit, store, and manipulate it at unprecedented rates and over great distances. Furthermore, society has recognized the utility of information and has developed major economic activities based on it, as well as greatly enhanced people's capacity to generate information through research. But the very existence of the developing countries indicates just how much knowledge the world still needs to narrow the gap between these nations and the developed countries. The task of the developed countries in achieving this is even more daunting than that of the developing countries because the consumption patterns of the industrial nations cannot be used as a model for the rest of the world.

THE GENERATION, TRANSMISSION, AND
DIFFUSION OF KNOWLEDGE

Every day some new knowledge is generated—for example, about nature, about how to create and use artifacts, about society itself—dispelling some uncertainties. At the same time, every day new uncertainties demand the generation of new knowledge. Thus findings about the existence of genes and their role in the replication of life raise a host of new questions: from how humans fit into the evolution of life, to the role of specific genes, to how genes can be altered to fight specific diseases and used to modify living organisms so that they better respond to human needs.

Today, it is impossible to quantify how much knowledge is being generated worldwide. Nor is it possible to know the extent of ignorance and of the unknown—of that infinite reservoir from which information is extracted and knowledge is shaped. More is known about certain kinds of knowledge, however, such as that of science and technology. For example, one can count the number of articles, books, patents, and the like in which portions of that knowledge are contained. (Other portions directly embedded in products or know-how are often harder to assess.) Another indirect measure for which rather precise information exists is the number and training level of knowledge personnel, such as researchers with doctorates.

As the end of this century nears, it is obvious that, in spite of many advances in knowledge, society still faces enormous uncertainties and ignorance in areas of immediate relevance to its well-being. Not only is it far from understanding many natural, social, and economic phenomena, but it also is far from knowing how to act on the basis of the knowledge and information it possesses. It does not know, for example, how to successfully educate all of the world population to respond to what is known about health and demography or natural hazards, or how to eliminate poverty.

Some of the impediments to dealing with these problems are in the social domain, while others are intrinsic to the genetic heritage of the human species, a heritage that is just beginning to be understood. But regardless of whether the necessary information is social, psychological, or genetic, one cannot act intelligently without it. This is a universal problem—for people, for organizations, and for countries, whether developed or developing.

The World Knowledge Infrastructure

The world knowledge infrastructure—the complex of systems for the generation, transmission, and utilization of knowledge—is large, ill-defined, and still in its infancy. Moreover, its growth has not followed a master plan; it often has occurred by the opportunistic aggregation of many elements, as in the case of many telephone networks. Nevertheless, some of the instruments needed to comprehend and envision ways to coordinate this infrastructure are now available.

For the *generation of knowledge,* or, in a narrow sense, the extraction of information from natural or human-made events and environments, the predominant organized players today are researchers and research institutions. The knowledge these players acquire is generally recorded and transmitted, albeit not universally received. Indirect measures of its magnitude show it is growing exponentially. In addition, a vast if poorly tapped body of valuable knowledge is being generated outside the research laboratories, by experience, by trial and error, or by chance. This largely grass-roots knowledge is recorded far less systematically, if at all; is transmitted haphazardly; and is impossible to measure even indirectly.

The instruments for the *transmission of knowledge* are myriad, ranging from schools, books, newspapers, and data banks to professional and scientific societies, conventions, information "highways" and postal systems, financial institutions, trade, technology transfer activities, and personnel exchanges. The International Congress of Scientific Unions (ICSU), with its well over 1,000 affiliated organizations from about 150 countries, exemplifies the global reach of these transmission instruments.

The World Wide Web (WWW), which encompasses the totality of network-accessible information, is another important and rapidly growing embodiment of human knowledge.[1] For developing countries, the Web, by exploiting the results of international collaboration, is an extremely important tool that gives them access to all kinds of information. Because it is "transparent" (democratic), modular, and accessible, the Web can enable a developing country to form its own network of information and make that information available to other developing countries and the rest of the world—and vice versa. But to benefit fully from the Web, a developing country must have in place the appropriate information infrastructure with channels of high bandwidth.

An increasingly promising aspect of the transmission of knowledge is the opportunity to use advanced telecommunications to operate laboratories and other research facilities at a distance—and eventually also to perform medical procedures. This not only can offer major savings to developing and developed countries alike, but it also can help scientists, engineers, and medical doctors from developing countries to work at the cutting edge of science, technology, and medicine. In industry, offices working in real time on the same problem or set of drawings in several locations that may be continents apart are already a reality.

Finally, the *utilization of knowledge*—putting knowledge to practical use—can involve potentially everyone everywhere. But to be effective, this again requires organizations and individuals capable and willing to act.

An integrated view of the generation, transmission, and utilization of knowledge—that is, the path from knowledge to action—should take into account the three distinctive interacting levels at which this process operates. At the top level are the international networks, agencies, and organizations that are primarily global transmitters of knowledge, such as the International Council of Scientific Unions (ICSU). Also at this level are found the international research institutes—such as the Consultative Group on International Agricultural Research (CGIAR) institutions that spearheaded the green revolution and the European Organization for Nuclear Research (CERN)—which will become even more important as individual countries find major research facilities increasingly unaffordable.

The mid, or country, level is where traditionally most scientific and technological knowledge has been generated—by universities, industries, and other institutions of a country. Transnational corporations reside at this level as well. Like many other institutions at this level, these corporations rely on the international networks of the first level for the transmission of information, and increasingly they also rely on their own secure point-to-point communications, bypassing public networks.

The lowest level is that of the grass-roots—individuals—where large amounts of knowledge are both generated and received. Pinpoint transmission of this knowledge among individuals is being enormously facilitated on a global scale by access to the networks of the top level. But because so much is now being received and generated at the grass-roots, people will require filters to reduce overload and to help to assess what is valid and essential. This is a particularly serious problem for developing countries. Markets, which are among the most effective instruments for the transmission of information, also operate at the grass-roots level.

The distinctions among the generation, diffusion, and utilization of knowledge are not sharp because the generators of knowledge also transmit knowledge (as in the case of World Wide Web, which was developed by CERN); the transmitters also utilize knowledge; and the utilizers can in turn generate additional knowledge. In spite of their academic tinge, the distinctions are useful, however, in helping to understand the workings, as well as the pathologies, of the knowledge infrastructure. Indeed, eventually they may help to determine the efficiency of the knowledge process—that is, the fraction of knowledge generated that is actually utilized.

For the three levels at which the knowledge process operates, the fundamental question is to what extent should a country or an institution contain the process at one level. For example, should the standards that guide the process be local or global? Should doctors and engineers be licensed to practice locally—or globally throughout the reach of international networks? Should a country developing a financial market think locally or globally? Similarly, should the language of instruction—that is, the communication of knowledge—be the local one or a world language? In each case the answers demand that one balance realism and a sense of vision and opportunity. But just one of many historical examples shows how difficult this is to achieve: the desperate and largely unsuccessful struggle of some of the rulers of the Ottoman Empire in 1800 to bring their institutions to a more modern international standard.[2] Current examples are the reluctance of some countries and institutions to linkup to the Internet, and the difficulties in reaching a consensus on how to achieve sustainable development.

How Does the Knowledge Infrastructure Actually Work?

The workings of the world knowledge infrastructure are still far from being well understood. Even in the absence of specific data, however, it is evident that much knowledge is not acted on for several reasons. First, only rarely is knowledge conveyed from generation to utilization through a direct pipeline. Much more often it is conveyed by a kind of general diffusion process, and usually only specialized gatherers, such as researchers or intelligence agencies, can harvest it effectively.

Second, the feedback from the users of knowledge (the entities and individuals that transform it into action) to the generators has not worked very effectively. Thus developing countries often receive knowledge that the industrialized world believes they should have, although it is not necessarily the kind of knowledge that the developing countries feel they need. This can be as much perception as reality, but the perception has led to accusations of cultural imperialism, and the reality has led to serious mistakes, such as the building of the ill-advised Artibonite dam in Haiti in the late 1950s. Today's worldwide information structure, made possible by advanced telecommunications, allows the feedback from need to generation of knowledge to be much faster and more effective. Furthermore, that feedback could operate more and more in both directions because the developing countries possess important elements of knowledge (such as native curative remedies and alternative medicine) that are in demand by developed countries.

Finally, many entities (such as countries, agencies, and individual companies) do not possess the mechanisms needed to receive information or sufficient information to act, or, even if well informed, they are not always capable of acting. Although this is particularly true for many developing countries, some industrial giants in developed countries have suffered grievously from some of the same failings.

On a global scale, the knowledge process has been successful in certain areas but not universally (Table 1). It has helped to contain some infectious diseases, to spread scientific and technical knowledge, and to create worldwide markets. But the process has worked at best only partially in the eradication of hunger and poverty and in the development of population policies, and it has worked very poorly in the avoidance of regional conflicts and genocide and in the preservation of ecosystems.

The Need for a Global Knowledge Strategy

The challenge is how to increase the effectiveness of the knowledge process—its ability to do what it is intended to do with as much economy of means

TABLE 1   The Information Infrastructure—Health Issues

Phenomenon	Information	Feedback Channels to Generation of Knowledge	Mechanisms to Receive Information	Mechanisms to Act on Information
Electro-magnetic radiation	Insufficient	Weak to strong	Strong (utilities, mass media, general public)	Strong to weak
HIV	Insufficient	Strong	Strong in some countries (medical establishment, media, and general public) Weak in some LDCs	Strong to weak
Cancer	Insufficient	Strong	Strong (medical establishment, mass media, general public)	Strong (medical establishment) to weak (prevention)
Cholera	Sufficient	Strong	Strong to weak	Strong to weak (developed countries vs. Rwanda, for example)
Famine	Sufficient	Generally strong	Strong (international agencies) to weak (some LDCs)	Often weak and uncoordinated
Health impacts of environmental degradation	Middling	Strong in developed countries (popular cause)	Strong in developed countries (mass media, general public); often very politicized	Generally slow; strong to weak (in LDCs)

NOTE: LDC = less-developed country.

as possible. This difficult goal cannot be achieved without a global knowledge strategy. But in spite of the work of many international agencies and the world-wide expansion of telecommunications and information, that strategy does not yet exist.

A global knowledge strategy should respond to two fundamental sets of questions:

1. What do people need to know to maintain and enhance their evolutionary advantage? What do they need to know as a species, as countries, as organizations, as individuals?

2. How do they build an effective global system to receive and act on that information?

In the context of these questions, the developing countries want to know not only how to implement long-term, fundamental changes, but also how to address immediate, down-to-earth concerns. For example, how can they initiate manufacturing activities even before the existence of a sufficient indigenous force of engineers? Or how can they improve agricultural production or health care before there are adequate laboratories or agricultural and medical schools? A series of pertinent and very specific considerations—both short and long term—are examined in the rest of this paper. Such an examination, however, cannot lose sight of the global context. If the human race is to have a future, the *global* improvement of economic and social conditions through better knowledge is imperative. This is a problem that both developing and developed countries must address jointly. It simply does the world no good in the long run if individual countries succeed in addressing their socioeconomic problems at the cost of neglecting such global problems as the growing depletion of the ecosphere or the potential for international conflict.

THE IMPORTANCE OF TECHNOLOGICAL
INNOVATION TO THE WORLD ECONOMY

Of the many facets of the process through which knowledge is generated, transmitted, and used, technological innovation is by far the most significant for the world economy. In his 1885 presidential address to the British Association for the Advancement of Science, Sir Lyon Playfair observed,

> France and Germany are fully aware that science is the source of wealth and power and that the only way of advancing it is to encourage universities to make researches and to spread existing knowledge through the community. . . . Switzerland contains neither coal nor the ordinary raw material of industry, and is separated from other countries that might supply them by mountain barrier. Yet, by a singular good system of graded schools, and by the great technical college of Zurich, she has become a prosperous manufacturing country. . . . The wealthy universities of Oxford and Cambridge are gradually constructing laboratories for science.[3]

Just about at the same time, German industry had been able to overtake that of France in the development of artificial coloring substances (Figure 1)—a dramatic demonstration of the impact of systematic research conducted in dedicated laboratories with full-time researchers. Such an example shows how relatively recent is a clear understanding of the importance of technological innova-

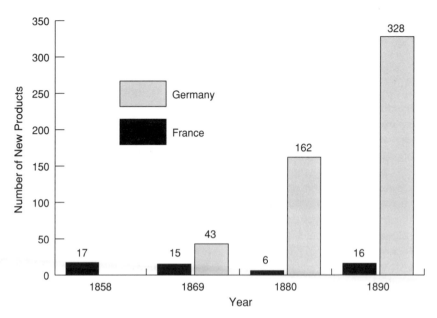

FIGURE 1   The development of artificial coloring substances in France and Germany, 1858-1890. SOURCE: Adapted, with permission, from François Leprieur, "La Formation des Chimistes Français au XIX siecle," *La Recherche* (June 1979): 732. © 1979 by Société d'éditions scientifiques.

tion to modern economies. Major technological innovations, however, have had a revolutionary impact on the world economy, directly or indirectly, throughout history—for example, the impact of scientific navigation initiated by Henry the Navigator in Portugal in the fifteenth century, or that of firearms and of the railroad. These impacts changed not only regional and national economies and global trade patterns, but also military and political balances and, in turn, the world economy. Examples abound. In the ancient world, an innovative naval technology enabled the Romans to defeat the Carthaginians, and such uncontested power made possible several centuries of peace and economic development in the Mediterranean region.

Today, the importance of technological innovation to the world economy is even greater. For the first time in history, the world has acquired the capacity to feed all its inhabitants, even if, paradoxically, for a variety of complex sociopolitical reasons, hunger still stalks the planet. And for the first time in history, thanks to technology, people have the potential to communicate with each other across the globe, to participate in a global marketplace, and to create a global "hyperintelligence" that could help to stem irresponsible population growth and enhance the prospects of the human race.[4] If the green revolution was a technological fix—one that will need to be performed over and over until population growth is brought under control—hyperintelligence represents a fundamental expansion of the social intelligence of the human species.

As economies have advanced, it has become increasingly clear that technological innovation, fueled by research and the generation of new knowledge, is now the major factor in increases in productivity and the *sine qua non* to guarantee a prosperous future. The enhanced ability to exchange information and transfer technologies will enable even the poorest developing countries to participate in the advances—not only economic but also social—made possible by technology. For example, the immense pain and inefficiencies associated with poor health could be reduced by a worldwide system of health care fostered by telecommunications advances. But the economic and social advances brought about by technology are not painless or without cost.

HOW DOES TECHNOLOGICAL CHANGE HAPPEN?

Technological change is a complex process. In the simplest terms, as for the airplane, radio, and nuclear weapons, curiosity or desire create a need; a need—or at times serendipity—creates an invention; and an invention creates a new technology. The new technology, if successful and useful, diffuses and overtakes the existing technologies—just as the railroads overtook canal transportation and motor vehicles and aviation overtook railroads. Eventually (and today very rapidly), the new technology is likely to spread through trade, the exchange of information, deliberate technology transfer, conquest (as in the case of the Mongols, who transferred knowledge and technology between East and West), or espionage.

The acquisition of new technologies does not occur, however, without a system in place that is intrinsically receptive or capable of being made receptive to innovation—for example, the Europe of the Renaissance.[5] At times, a system is forced to be receptive such as in the Russia of Peter the Great and later of Lenin and Stalin. Today, the existence of a national or international system of innovation is recognized to be a key factor in technological advances.[6] Among the many conditions that make a system receptive to inventions and innovations, investment in the scientific and technological infrastructure and associated human resources is paramount. But because many developing countries lack such an infrastructure, they must try to operate with what they have, while patiently and systematically developing the components of the technological infrastructure required to advance.

An effective technological infrastructure enables a country to generate and utilize knowledge. It includes an educational system that encourages creativity and the pursuit of scientific and technological knowledge at all levels; an educated and skilled work force; a network of outstanding research laboratories; measures and standards laboratories; critical machine shops; linkages between researchers or inventors and the potential utilizers of the knowledge they generate; a legal system to protect technological innovation, whether indigenous or imported; a financial system to invest in innovations; a fiscal system to encourage innovation; and a general population that is technologically literate and receptive to innovation.

The last factor is often not given enough weight, yet history has proven its importance. In the second half of the eighteenth century, the *Encyclopédie* of Diderot and d'Alembert provided an increasingly interested general public with a vast array of technical and scientific information. From the seventeenth to nineteenth century, the ingenuity of a number of American inventors, who though often not formally educated were technologically literate and operated in a very receptive milieu, provided the new country with a variety of important technologies, from gin mills, firearms, steamboats, and bridges, to Henry Ford's automobiles and the Wright brothers' airplanes. In Japan, from 1853 to the battle of Tsushima in 1905, the transformation from a feudal technology to a very advanced modern technological level stemmed largely from a disciplined, educated work force that was open, as it continues to be today, to technological innovation.

Among many developing countries, particularly the poorer ones, widespread technological interest has yet to emerge, hindering in many ways the process of technological development that has been so successful in Europe, America, Japan, and parts of Asia. If technological innovations are to occur, those developing countries must emphasize technological literacy in schools and encourage it among the entire population.

Three other factors also are essential to innovation. First, leaders must be well educated and seriously convinced of the importance of technological innovation. And they must be prepared to act. Undoubtedly, one of the factors contributing to the success of Taiwan in recent years is the number of Ph.D.s in leadership positions, including current cabinet members and the prime minister.

Second, research laboratories must focus on the needs of the country. For a developing country this means a strong emphasis on applied research and development, but windows on basic science, by maintaining a core of basic researchers who can follow and participate in world advances in science, are very important and should not be overlooked. Research laboratories and universities also should avoid excessive compartmentalization. The scientific, technological, and social challenges of the developing countries, like those of the rest of the world, require interdisciplinary interaction: materials engineering needs to interact with biology, computer science with linguistics, medicine with sociology, engineering with economics and the law, and so on.

Third, developing countries need to pool their scientific, technological, and educational resources with those of neighboring countries to create critical masses of resources that are beyond the capabilities of individual countries. Europe has been doing this with considerable success, as exemplified by the Concorde, the

Airbus, its space program, and a large number of European Union programs, from Eureka to Erasmus, that encourage intra-Union cooperative science and technology projects. Most developing countries, of course, are still a long way from being able to emulate these programs, but they would greatly benefit from regional technological cooperation. If that cooperation is pursued judiciously, starting with specific projects and extending to joint technological and industrial policies, its benefits are bound to far outweigh those of competition among neighbors. But participants in such arrangements must guard against two traditional dangers: the possibility that a regional center will siphon off scarce talent from the participating countries, and the tendency of a center to serve best the needs of the country in which it is located. Here again technology can help by making possible the creation of "distributed" or "virtual" centers, most of whose personnel are located in their own countries. The Third World Academy is attempting to overcome the subcritical mass of outstanding scientists in most developing countries by operating as an academy of sciences for all of the Third World.

THE COMMERCIALIZATION AND
GLOBALIZATION OF INNOVATION

Some innovations that fill in very obvious needs—such as the radio, automobile, airplane, bicycle, television, or x-rays—become easily commercialized and spread globally. Even so, there are almost always struggles. Radio was not developed in the country where it was invented; the significance of airplanes as transports or offensive weapons was not envisioned at the beginning; the naval screw propeller took a long time to prevail over paddle wheels. Thus often even the most useful technological innovations have to overcome the hold of conservative older technologies. Innovations that do not respond to an obviously perceived need or desire, such as the Walkman, require even greater foresight and perseverance in their commercialization. And then, of course, there are those innovations such as nuclear weapons that one struggles not to see globalized.

Each element of the relation technology-products and production-market (TPPM) plays an important role in the commercialization and globalization of technological innovation. But, unfortunately, many developing countries, in addition to being intrinsically weak in technology, are unrealistic in the selection of marketable products, inefficient in their production, and naive in marketing. In addressing these problems, a developing country, like any other country, must consider a number of issues that go well beyond the category of "technology transfer." Indeed, technology transfer is only one of the instruments required by a national system of innovation—a system that needs to look just as much inward for innovation as outward. Altogether, these instruments must include or consider:

- *The structure of industry.* The size distribution of industry in a developing country is frequently very skewed, with a preponderance of very small companies, very few mid-size companies, and only a few dominant large companies. The very small enterprises, based perhaps on the work force of a single family, do not have the resources and know-how to benefit in a major way from the results of research, let alone do research. Thus research and development are virtually nonexistent in the industry of many developing countries, except perhaps in some large companies. Nor is there a productive "industrial ecology," such as a system of relations between small innovative companies and larger companies that are the recipients of innovation. Although this kind of system has succeeded in the United States, other systems prevail in other countries, where small companies usually serve as suppliers to big companies, which are the main source of innovation. Under these conditions, universities often become even more important as sources and transmitters of knowledge.

• *Standards* (agricultural, industrial, telecommunications, accounting procedures, etc.) and the quality of products and of the production process, including services. The keys to commercialization and globalization, standards and quality are still far from well developed in most developing countries.

• *Telecommunications infrastructure.* This infrastructure is very crucial but usually is very weak. Yet, thanks to satellites, developing countries have the opportunity to leapfrog several stages and immediately acquire advanced telecommunications capabilities.

• *Information databases* for science, technology, management, and world trade (markets, suppliers, etc.). Such databases are usually very weak, but they can be built more rapidly today.

• *Communications infrastructure* (roads, airports, navigation). The inefficiencies of this infrastructure are costly, causing, for example, rotting of agricultural products, larger-than-needed inventories, and slow payments. Furthermore, the dreams of intracontinental networks of highways or railroads with their associated major benefits survive today only in Europe, Australia, North America, and parts of Asia.

• The *"soft" service infrastructure* (accounting, patent lawyers, other legal services, etc.). Components of this infrastructure must be oriented toward the needs of technological innovation and global markets.

• *Appropriate educational system.* The educational system is the key instrument through which information is being generated and transmitted to those who will use it (ideally, all citizens) and who will, in turn, generate new information. A frequent mistake made by developing countries is that of following uncritically the examples of developed countries and perpetuating them by counterproductive mechanisms of faculty selection and career advancement, based, for example, on the production of papers rather than on the ability to teach effectively and to focus on national problems. In a number of developed countries, faculty interactions with industry are limited, and the duration of an education is much too long, bringing to the job market students who are too old and have lost some of their spark and flexibility. More important, the engineering schools of developed countries, which are attended by many students from developing countries, do not prepare these students fully to tackle development problems. These schools might, for example, educate a cadre of "development engineers" who are able to work not only with the traditional large public projects, but also with private enterprise. This new breed of engineers, as the technical strategists of development, should possess a keen sense of how systems work, as well as broad sociotechnological knowledge. They should know, for example, how to identify a development problem; how to mobilize the forces and the resources needed to solve it; how to address critical infrastructure needs; and how to help to generate jobs by giving small companies the tools they need, from expert advice to access to information and research. Groups of internationally oriented specialists, who know markets and how to reach them, and who understand the competition and the viable niches for the products and services of a developing country, should be trained as well.

• *Organization.* In most developing countries, the organization of science and technology in government, science and technology policies, industry organizations, university-industry mechanisms, and peer review mechanisms, among other areas, are seldom given the priority or addressed with the depth they demand. Science and technology policies, however, cannot stand alone; they must be connected to economic, financial, and social policies. The developing countries also need to learn the techniques of science and technology policy research, which will serve as the basis for science and technology policy decisions.[7]

• *Investments in "technological observatories," joint ventures, and other arrangements.* Developing countries might consider locating "observatories" in other, more technologically advanced countries to help to assess future scientific and technological directions and how to profit from them. These observatories, which are used effectively by major companies in developed countries, range from laboratories around universities and other environments that produce relevant knowledge, to investment through venture capital funds in emerging technological companies of promise.

Joint ventures or other arrangements could further extend the technological observatory concept, and the possibilities are intriguing, even if untested. Indonesia, for example, is investing in an advanced production facility in a technologically advanced country to serve as a training ground for developing country personnel, as an originator of new designs that bring together the experience of both countries, as a technological observatory, and as a port of entry into the global market.

• *Encouragement of investments from abroad.* Such investments bring with them new technologies that can in turn become the basis of indigenous industries.

• *Creation of clusters of research-production-marketing activities,* such as applied research parks or university-industry parks. These clusters will facilitate the connection of research to production and the marketing of the results. Rather than dispersing assets, the parks offer synergistic concentrations of knowledge, workers, and facilities.

• *Focus on production efficiency.* The advantage of cheaper labor is lost if a developing country suffers from inefficient production. Increased production efficiency requires the introduction of automation, preferably by a cadre of efficiency specialists capable of turning around an industrial company or service organization. Although the route to economic well-being for a developing country is through higher pay for its workers, this goal must be pursued gradually to enable the country to maintain its cost advantage while progress is achieved in other areas.

• *Encouragement of markets and the private sector.* Markets generate and utilize an immense amount of information and have been crucial to technological innovation in the developed world. The New York Stock Exchange is too complex and sophisticated to serve as a model for developing countries, but simpler electronic markets are made possible by computer technology and the new telecommunications networks.

As for the importance of the private sector, over the past decade six out of seven new jobs in the United States have been generated in the this sector. Moreover, the rate of job creation in the United States has been higher than in Europe, where the majority of jobs has been generated in the public sector.[8] To achieve a high rate of job creation, the developing countries may need to enact policies that enhance the dynamism of their economies, more than emphasizing only job preservation and creation. This is the kind of hard-won knowledge that can help to guide the strategies of the developing countries. Information from both developed and developing countries and candid appraisals of the results of different policies can enhance the knowledge of how the latter can generate jobs and balance economic objectives with social compassion and environmental preservation. This is a particularly complex challenge for developing countries because with their widespread poverty they cannot adopt uncritically economic development measures that might place their citizens at even greater risk.

• *Intelligent choices of directions.* Although intelligent choices in technological development are very important, much too often countries—and not just developing ones—cannot discriminate and focus appropriately, particularly those countries with weak, market-driven private sectors. The key questions are how to add, through technology, more value to the products and services of each country and how to create new industries. The search for the best approaches to adding value requires systematic exploration of all potential technological opportunities, whether in agriculture, forestry, marine resources, manufacturing, or services. These opportunities might range from preserving products, to industrializing the harvesting of resources (for example, aquaculture), extracting derivative products (such as drugs, vitamins, and glues), devising new uses for products (such as wood prefabs and glued wooden bridges), and expanding the services offered by tourism and other activities. For energy and mineral resources, opportunities may lie in refining within the country rather than shipping abroad.

• *Understanding the "simplicity threshold"*[9]—the threshold that separates simpler products or processes from more complex ones that require high sophistication in design and production—and how it relates or can relate to the industrial infrastructure of a country. Such an understanding is crucial in making intelligent choices about the creation of new industries. Today, 75 percent of world trade—such products as motor vehicles, computers, telecommunications equipment, and aircraft—are above this threshold. The products below the threshold—such as petrochemicals or simpler industrial components—constitute a shrinking percentage of world trade and have smaller value added. Again, in the area of processes a growing portion of products in the world market stems from complex processes above the threshold.

But such percentages do not mean that the world will not continue to need the many basic products below the simplicity threshold. Rather, it means that to advance, a developing country must find ways to add value to those products and should not rely exclusively on them. The move from simpler products and processes, including services, to more complex ones depends critically on knowledge and on the ability to create learning environments. In these new organizational environments, the acquisition of knowledge and feedback from experience are viewed as the *sine qua non* for survival and progress. But their creation will require profound transformations of often ossified or ignorant bureaucracies.

In endeavoring to pass over the simplicity threshold, a number of former developing countries, including Taiwan, Korea, Singapore, and Malaysia, have succeeded in going down the path from selling foreign products internally to manufacturing parts for those products, to manufacturing the entire product. In the future, however, that path is not likely to be as linear as in the past. The industry of a developing country will have to learn not only how to supply parts and components to companies abroad, but also how to master in turn the outsourcing process by integrating products and services acquired from abroad. Furthermore, if a country has a large cadre of well-educated specialists, it also can supply the world market with advanced labor. For example, today India, Pakistan, and Russia provide software and software designers for American and European companies, an activity greatly enhanced by telecommunications.

During its development, Japan followed a different, less common, and ultimately, in terms of an open global economy, a less desirable path by typically importing a prototype or concept, followed by retroengineering and full-scale, continually improved product manufacturing. This was possible only because Japan had a highly skilled technical work force and strong protectionist policies. It was facilitated by a large internal market that could subsidize exports through higher internal prices and could absorb the initial run of products so that they could be tested before reaching the world market. The size, current or potential, of the internal market is always a most important strategic determinant, regardless of the industrial strategy adopted. Thus Indonesia, with a population of 200 million and a rapidly advancing—even if still extremely low—gross national product (GNP) would follow a strategy quite different from that of Cuba.

POSITIVE AND NEGATIVE EFFECTS OF TECHNOLOGICAL INNOVATION ON DEVELOPING COUNTRIES

Technological innovation has a tremendous impact on developing countries—on their agricultural production, health care systems, education, and work habits. These impacts in turn greatly affect family size, urban migration, everyday life, the condition of women, and other social concerns. For example, although improved conditions for women are, in general, still much too slow in coming, women are training to be astronauts in Indonesia, studying engineering in Libya, and holding increasing numbers of professional positions in many developing countries. The close correlations observed between higher literacy rates and lower fertility rates also are remarkable.[10]

The political impacts of technological innovation have been equally significant. Telecommunications helped the coming to power of the Ayatollah Khomeini in Iran and the development of political consciousness in disenfranchised populations. The dismantling of the Soviet Union in 1991 was triggered by an increasing inability to compete technologically with economies that had reaped the benefits of advanced systems of innovation.

Different developing countries with differing value systems will inevitably see these impacts in different lights, but the impacts cannot be ignored. The inescapable fact is that the introduction of new technologies brings with it irreversible social and political change. Once some technological innovations are introduced, and with them a glimmer of a different future, a country, for better or for worse, is never going to be the same. Even strong authoritarian regimes (such as China or the former Soviet Union) cannot stop the process of change.

Some of the impacts of new technologies can be negative. When they occur, such impacts must be recognized and mitigated, even if they are often hard to foresee and even harder to modify. This is the responsibility not only of the developing countries, which can be easily tempted to accept new technologies uncritically, but also—and above all—of the developed countries.

One negative impact might be *loss of local economic autonomy*. In a region producing for the world market, fluctuations in that market or competition from other regions can be devastating. Thus it is imperative that a developing nation differentiate its products from others, as well as seek efficiency, quality, higher value added, and regional cooperation. Technological innovations that render the industrial products of a particular developing country obsolete can come on the scene very rapidly. The market also can be capricious, as in the case of fashion, or affected by new concerns, as in the case of the environment. These are all factors that can put at risk a country that does not try to anticipate change, or that relies too much on an economy based on few products.

Yet another negative impact is an *extraordinary dependence on supplies from other parts of the world* such as fuel, machinery, information, and personnel. The effects of the recent embargoes on Haiti and the isolation of Cuba are examples of how desperate the situation can become when those supplies cease to be available. And in Africa, the Sahel suffered widespread devastation when fuel became scarce and prohibitively expensive. Of course, no country in the world can be an island unto itself, but developing countries with limited resources are particularly vulnerable.

Increased use of resources—an inevitable correlate of economic development—leads to environmental depletion and increased pollution and waste. The tragedy of many developing countries is the merciless exploitation, borne out of desperation, of their natural resources and the devastation of their environments. Abundant, inexpensive labor should make it possible to create new, appealing environments offering a refreshing contrast to those of highly industrialized countries. This potential advantage should not be lost if a developing country is to provide a higher quality of life for its citizens and if it seeks to attract tourism as well as commercial, service, and industrial nonpolluting operations from developed countries.

Unrealistic expectations and aggravated internal inequalities also are negative side effects of technological innovation. Once a country embarks on the path toward technological development, the expectations of its population almost inevitably exceed the ability of the country to bring social and economic advances to all segments of its population. The result is the potential for unrest and the persistence of poverty in many regions. For example, from 1980 to 1990 in Latin America, families in poverty (defined as the percentage of people lacking the income for a minimum level of food, shelter, health, and educational services) increased from 35 to 39 percent. Poverty is projected to remain at the 38 percent level to the year 2000, in spite of positive increases in the gross domestic product (GDP) from the late eighties to today.[11] In China, the economic quasi-laissez-faire policy of the government has encouraged a great deal of technological and economic development, but at the price of great imbalances in the affluence of different segments of the population. Even in Europe and the United States, these imbalances have been very hard to overcome. The fact is that in a world society in which the traditional source of employment for the uneducated—manufacturing—is becoming highly specialized, the key to social and economic advances is knowledge. Thus inequalities in access to information networks and in the ability to use information are particularly serious and must become a major concern of the developing countries and the rest of the international community alike.

The breakup of traditional culture is perhaps the most distressful aspect of technological change since such change can have a much greater impact on developing countries than on most developed countries, where the changes have occurred more gradually and are much more the result of indigenous technological developments. In a developing country, technological change arrives almost by definition from the outside world in the form of full-fledged products that sweep the country without giving social structures and mores sufficient time to adapt. Thus the developing countries need assistance in fortifying key elements of their culture before they are irretrievably swept away to the loss not only of the developing countries but also of the rest of the world.

FORECASTING AND ASSESSMENT IN INVESTMENT DECISIONS

The fundamental question when making an investment decision is: Will the investment achieve its goal? Difficult for any country, this question is particularly so for a developing country that is unlikely to have sophisticated mechanisms for assessing risks and benefits. In any major investment and borrowing decision, a developing nation may place at risk its very future.

Risk is a multifaceted sociotechnological problem. Some risks are purely technological—for example, an intensive irrigation project may cause salination of the soil or saltwater intrusions in the water table, a reservoir may become silted too rapidly, or a new industry may turn out to be technologically noncompetitive.

Financial risks could take the form of an investment that may not be repaid, or the rate of return fixed at the outset may become less desirable over time than alternate investment opportunities. Moreover, a recipient country may continue to borrow from other sources and accumulate an irrationally structured debt that is hard to assess but a serious risk.

There also are the social risks that a project (or the fiscal conditions associated with its financing) may turn out to be counterproductive. For example, the required fiscal discipline may place the leadership of a country in an impossible position and cause drastic political changes (which has happened repeatedly in South America). Or the productive capacity may shift counterproductively from the countryside to the cities. The associated increase in urban population entails all kinds of risks: invasion of fertile land (as in Cairo), greater exposure of large population conglomerates to disease and natural hazards such as floods or earthquakes (but also to better educational and economic opportunities), changes in the social support system from extended to nuclear families, crime, irreducible poverty, and greater opportunity for social upheavals. Because these kinds of risks are so difficult to assess, some international assistance agencies have simply given up trying.

A particular set of socioeconomic risks stems from the difficulties in forecasting demand. While it is clear, for example, that energy demands, like those for water, tend to increase with population and GDP, they are more difficult to predict quantitatively because they are affected by increases in production efficiency, by emerging new technologies, and by possible changes in the pattern of energy use. Energy demands also are affected by international cartels and the cost of energy supplies, and by concerns about the environment that were either underestimated or not foreseen when a power plant was constructed. This occurs, of course, in both developing and developed countries, as exemplified by the reduced emphasis on nuclear energy in Germany, Italy, and the United States.

Given these risks, what should a developing country do? Primarily, it must develop strong mechanisms of technology assessment and risk evaluation. For any country, technology policy decisions are never isolated from political considerations. But the existence of a creditable capability for science and technology assessment and for science and technology policy is essential to survive in a knowledge-oriented world. That capacity must be honed by experience and frequent contacts with corresponding entities in other countries and with such international organizations as the World Bank.

Technology assessments should not be only retrospective; they also should operate in a feed-forward mode to determine foreseeable future impacts, the probability and magnitude of future risks, and possible corrective measures to reduce them. But in undertaking its assessments and risk evaluations, a developing country should not operate independently of an organization like the World Bank that endeavors to assist it and of other countries that are potentially relevant. In turn, the Bank, however strong its technical resources, can benefit from the information obtained from such institutions as the National Research Council, with its capacity to mobilize in a multidisciplinary fashion a large number of highly qualified scientists, engineers, economists, and health care personnel.

Two principles are particularly important for a developing country embarking on technological forecasting and assessment. First, it must avoid being captured by projects that may appear glamorous but do not respond to the urgent needs of the country or are unrealistic, such as being far above the simplicity threshold. Examples abound not only in developing countries but also in the industrial countries, the main difference being that developing countries cannot afford to invest very scarce resources poorly. Second, regardless of what direction the future may take, close attention to the connection between technology and socioeconomic developments is imperative. Failure to see the many possible pathologies of that connection is a major cause of breakdowns in the use of technology for development.[12] A sociotechnological factor of considerable importance, among many, is the different time constants of technological development and political life and associated different levels of knowledge and experience. Political changes occur both in the developing countries and in the more developed countries with which the developing countries need to interact in order to sustain their technological development. But the success of any technological development depends on continuity of conception, creation, and operation. If the political life span of persons involved in the technological process is much shorter, the learning curve must be started over and over again. The World Bank, with its inherent greater continuity, can perform a most needed balancing role in this regard.

WHAT ARE THE FUTURE COMPARATIVE ADVANTAGES FOR TODAY'S LABOR-SURPLUS DEVELOPING COUNTRIES?

As the world economy becomes ever more knowledge-based, it is clear that the value added made possible by information (by science, advanced design concepts, technology of intelligent materials, automation, software, sensors, advanced services, new medical concepts, etc.) is propelling today's economically more advanced countries. Thus to close the gap with these countries, developing countries cannot continue to rely over the long run on cheap labor. While a developing country cannot adopt the capital-intensive strategies of developed countries, it should not see itself stuck hopelessly with a labor-intensive strategy; a knowledge-based strategy offers it the opportunity to emerge from such a situation. The labor cost gap with the more advanced countries will close slowly, however, in spite of the exceptionally fast rates demonstrated by such countries as Taiwan or Korea. This will give many developing countries some breathing space, but they cannot stand still lest they fall further below the simplicity threshold. Even the poorer developing countries will need to gradually introduce computers and automation, enhanced educational systems, and appropriate science and technology policies to begin to transform their work forces and economies.

It is imperative as well that they add value to the fruits of their labor through an appropriate selection of products and technologies and an effective mix of the different factors of production. For example, the well-trodden path from the production of parts to that of complete assemblies offers the opportunity to add further value to those assemblies through, for example, software. But it would be foolish for a developing country to believe that it can utilize its labor cost advantage for the creation of an automotive industry, or any other industry far beyond the threshold of product and production simplicity. Opportunities in the service sectors that are labor-intensive in developed countries also should not be overlooked. For example, care of the world's increasingly large number of aged—care that cannot easily be supplied by relying only on technology—could be among several potential niches for developing countries. The production of custom-made, labor-intensive artifacts will remain a niche for a long time, but because of the growing saturation of consumer markets, this is not an activity that can offer employment to large numbers of workers unless it is coupled with ways to add further value to the artifacts. In short, a systematic approach is needed in order to find and exploit areas of opportunity.

CONCLUSIONS

More than ever before, the survival and progress of nations and their citizens depend on their abilities to generate and utilize knowledge. For developing countries, economic and social advancement will be achieved by learning to tap the global system for the generation and transmission of knowledge and to develop effective mechanisms for transferring knowledge into action. This process should not be a one-way street, however. Global networks also can unlock and distribute the knowledge that the developing countries possess and that the rest of the world needs—embedded, for example, in the genes of their plants and animals or associated with time-tested healing practices. One of the most important elements of knowledge for developing and developed countries alike is gained through sharing and comparing their experiences—successful or not—with technological, economic, and social development.

The best hope, then, of narrowing the gap between the developing countries and the more developed countries is a knowledge strategy. Such a strategy will enable a developing country to bypass steps in the development of technological systems that historically have had a long gestation—from telecommunications and transportation to medical services and the applications of genetic engineering. But the benefits of not having to repeat the entire learning curve for many technologies will be illusory if a developing country does not create the appropriate infrastructure to use knowledge effectively and to participate in the generation of new knowledge.

Obviously, each developing country must define the knowledge strategy that best suits its needs. The greatest challenge to the developing countries will lie in developing a new kind of sociotechnological knowledge that will enable them to advance without following the unsustainable model of today's industrial nations. The problem, of course, is that most developing countries are poorly equipped to develop and apply this new knowledge. Thus they need to work imaginatively, in close collaboration with international institutions and with developed countries, which by now have acquired a broader perspective on their own development and its impacts.

In formulating their strategies, developing countries also must consider how they can position themselves with respect to the simplicity threshold, how they can define the most desirable proportion among the different factors of production, how they can best meld indigenous creativity and innovation with the import of products and ideas, and how they can balance technological advances with social health—all of which will spell the difference for a developing country between success and inability to take off. And across the globe, a pivotal element of any knowledge strategy must be stronger national knowledge infrastructures and their global integration. The time has come to determine how this will be achieved.

NOTES

1. Brian Hayes, "The World Wide Web," *American Scientist* 82 (September-October 1994): 82.

2. See, for example, Andrew Wheatcroft, *The Ottomans* (London: Viking/Penguin, 1993).

3. George Basalla, W. Coleman, and R. H. Cargon, eds., *Victorian Science* (Garden City, N.Y.: Anchor Books, 1970).

4. George Bugliarello, "Toward Hyperintelligence," *Knowledge: Creation, Diffusion, Utilization* 10 (September 1988): 67-89.

5. John Hale, *The Civilization of the Renaissance in Europe* (New York: Atheneum, 1994).

6. See, for example, Jorge Niosi et al., "National Systems of Innovation: In Search of a Workable Concept," *Technology in Society* 15 (1993): 207-227; and Nachoem M. Wijnberg, "National Systems of Innovation: Selection Environments and Selection Processes," *Technology in Society* 16 (1994): 313-320.

7. See, for example, Edward Wenk, Jr., *Making Waves—Engineering, Politics, and the Social Management of Technology* (Urbana: University of Illinois Press, 1995), 11.

8. Amity Shlaes, "Germany's Chained Economy," *Foreign Affairs* 73 (September-October 1994): 109-124.

9. Robert W. Rycroft and Don E. Kash, "Technology Policy in a Complex World," *Technology in Society* 16 (1994): 243-267.

10. Wade Roush, "Population: The View from Cairo," *Science*, August 26, 1994, 1164-1167.

11. Nathaniel C. Nash, "Latin Economic Speedup Leaves Poor in the Dust," *New York Times,* September 7, 1994, 1, 14.

12. George Bugliarello, "Technology Transfer: A Socio-Technological Paradigm," in *Development and Transfer of Industrial Technology,* ed. O. C. C. Lin, C. T. Shih, and J. C. Yang (Amsterdam: Elsevier Science, 1994).

What We Know and Do Not Know about Technology Transfer: Linking Knowledge to Action

HARVEY BROOKS

Professor of Technology and Public Policy, Emeritus, John F. Kennedy School of Government, Harvard University

Technology transfer is a way of linking knowledge to need. In fact, a continuum exists between the assimilation and creation of new knowledge, although in practice it is customary to think of technology transfer and research and development as separate activities. Such a continuum is important, because there is a tendency, especially in the developing countries, to think of technology transfer in minimal terms—for example, the transfer of the minimum amount of know-how needed to enable a specific manufacturing plant to produce certain products. Too many politicians and managers in developing countries view technology as a package that can be bought "off the shelf" and become immediately useful. But rightly conceived, technology transfer is a process of cumulative learning, in the same sense that research and development (R&D) are a form of cumulative learning.

Thus neither technology transfer nor R&D are synonymous with innovation. Each is just one element in a much more complex process that is better described as "social learning," or, more accurately, as "sociotechnical learning" since technical knowledge, organizational knowledge, and new relationships among people inside and outside an organization have to be absorbed and "internalized" in groups of people before an innovation or even a production plant can be sustainable. In this sense, there is much less of a difference than generally supposed between a production system that is new to the world and one that is merely new in a particular sociotechnical context characterizing a particular manufacturing site and market.

For this reason, the process of creating a production system at a new site always can be considered, at least in part, an innovation because, though not new in the world, it is new in its particular social-economic-political context. The process itself is thus fundamentally similar in developed and developing countries, whether one is replicating something that has largely been done before elsewhere, or doing something that has never been done before anywhere. It is the absorption of knowledge into a system of design, production, and interaction with clients or customers that is critical; the novelty of the knowledge in context is really the critical variable in the process.[1]

One implication of this viewpoint is that "know-why" is often as important as "know-how" in the process of absorption because only if the reasons for particular technological choices are understood can the transferred knowledge be built on in a cumulative manner and the processes continually improved and made more efficient. Thus Richard Nelson defines innovation as "the processes by which firms master and get into practice product designs and manufacturing processes that are new to them, whether or not they are new to the universe, or even to the nation." Furthermore, he defines technology not only as "specific designs and practices" but also as "generic understanding . . . that provides knowledge of how [and why] things work" and of "what are the most promising approaches to further advances, including the nature of currently binding constraints."[2]

This view of technology transfer begins therefore with the premise that the conditions for successful technology transfer are basically similar in developed and developing countries and differ only in the fact that in developed countries much of the transfer takes place between research and development organizations, whether inside or outside the firm, and other parts of the firm, while in the developing countries technology is usually transferred from outside both the country and the firm. In each case, a "culture gap" must be bridged, although in the case of developing countries it is likely to be much larger than within or among developed countries. Nevertheless, the basic challenges of organizational absorption are quite similar in kind if not in degree.

THE BASICS OF TECHNOLOGY TRANSFER

The process of technological innovation can be described as one of matching solutions in search of problems to problems in search of solutions. Solutions in search of problems are mostly produced by corporate and academic research laboratories and other forms of organized research and development, or, in the case of developing countries, are available somewhere in the industrialized world. Problems in search of solutions are what industry, society, and design engineers encounter in practice. Solutions in search of problems usually have proved to be the most efficient way to create and package knowledge for ready communication, but in that form it is usually not most easily used by appliers of knowledge. Improving the "impedance" match between these two forms of knowledge is a primary task of technology transfer.

Until quite recently, one problem of U.S. technology policy was its implicit overemphasis on R&D as synonymous with technological innovation and as the key focus of technology policy. The mission of technology policy was thought to be to get R&D right, on the assumption that everything else would take care of itself more or less automatically. Some of this same attitude also has tended to govern development strategy: the initial technology choice is the only critical factor rather than the entire process by which the results of this choice are ultimately internalized in the overall organization of the firm and its work force.[3]

But even in the developed world R&D represents only a small fraction of the total investment needed to get a new technology to market—about 10-15 percent on average according to the famous Charpie report.[4] The other 85-90 percent is usually referred to as "downstream" investment in design, manufacturing, applications engineering, and human resource development—the last involving a great deal of hands-on training in the context of actual operations. Although not all of these activities are purely technical (they are better described as "sociotechnical" because they involve changes in organization and human relations as well as technology), they do require a heavy commitment of experienced technical personnel. Indeed, about 65 percent of all professional scientists and engineers in the national work force in a typical industrialized country are not engaged in R&D at all but in a broad spectrum of downstream activities. Much of the technical activity from which "economic rents" may be derived resides not in the R&D but in these downstream activities, and this becomes increasingly so as technology becomes more "science based." But because of the way in which statistics on technical activity in the United States have been collected, much more is known about the nature, quality, and content of R&D than about the quality and value-added of the downstream technical activities. The same is even truer in most of the developing world, where these downstream elements of the "technical capacity" of a firm represent an even larger fraction of the total effort.

Over the last 30 years, much research has been conducted on the features of the innovation process that lead to commercially or operationally successful technological innovation. This research began with the chemical and scientific instrument industries in the 1960s and since then has been extended to the machinery and electronics industries. According to Christopher Freeman,[5] who summarized the main conclusions of this work at a 1990 Montreal conference on networks of innovation, there are six sources of innovative success:

1. Understanding user needs and establishing user-producer networks, as well as attempting to understand the special circumstances and needs of potential users of products and processes. Strong and continuing user-producer linkages are vitally important.

2. Coupling development, production, and marketing activities, and considering manufacturing and marketing requirements at an early stage of development. Ongoing technology assessment should be carried out based on such considerations and on monitoring technological developments outside the firm.

3. Linking up with sources of scientific and technical information and advice outside the innovating organization. Outside networking is essential even with strong in-house R&D. Inside and outside sources of information are complementary, not alternatives. Internal R&D should tap into external science and technology networks in order to facilitate the generation of new knowledge internally.

4. Concentrating high-quality R&D resources on the innovative project. A critical size of R&D effort is necessary to realize internal goals and to match the activities and investments of competitors. A strong in-house technical effort is essential for understanding the significance of new technical developments outside the firm.

5. Seeking a relatively high level of performance of some relevant basic research within the firm, which usually correlates strongly with successful and timely innovation largely because it tends to enhance early awareness of technical developments that might affect the evolution of the innovation.

6. In the firm, an entrepreneur/innovator is generally characterized by high status, wide experience, and seniority within the organization. Top management has a high degree of commitment to the success of the innovation and performs a network coordination function both inside and outside the firm, serving as its principal innovation champion.

Almost all of these items are applicable to developing countries. Although R&D resources (item 4) as such are less important, the equivalent of R&D is high-quality training applicable to the particular products and processes involved. Such training, both generic and specific, should be an important part of any technology transfer package.[6] The entrepreneur and the entrepreneurial spirit (item 6) are equally essential in developing countries, but they are often more foreign to the indigenous culture than is the case in industrialized countries.

One of the major trends in the developed countries over the last 30 years has been the increased importance over time of the sources of technical information and ideas that originate outside firms, including the growth of institutional alliances and "innovation networks," frequently crossing national boundaries. Many of these structures are ad hoc and temporary, formed for particular innovations or production plants. This results in a complex intermingling of competition and cooperation, with some firms cooperating in selected projects while at the same time competing in other areas. Some of the varieties of institutional interdependence and cooperation are:

1. Joint ventures and joint research corporations
2. Joint R&D agreements among firms
3. Technology exchange agreements
4. Direct investment motivated by technology factors
5. Licensing and second-sourcing agreements
6. Subcontracting, production-sharing, and supplier networks
7. Research associations

8. Government-sponsored joint research programs

9. Computerized data banks and value-added networks for S&T exchange

10. Informal and only partially sanctioned information sharing among technical people in competitive firms.

The kind of exchange described in item 10 is often not strongly opposed by management because there is an implicit expectation of future reciprocity that makes the joint gains from such cooperation exceed the possible competitive losses in the long run.[7]

In short, as product design and production technology have become increasingly science-based, know-how has tended to diffuse more and more rapidly throughout the world technical community. As a result, the competitive advantage from which economic rents can be derived depends more on the downstream details of implementation and less on the novelty and originality of the basic technical idea or generic design. Moreover, more rapid diffusion greatly narrows the window of opportunity available for getting into the market with competitive technology and products, thereby reducing the chances of recovering the initial costs of innovation or investment in new products and production systems within this window. At the same time, the chances of being outclassed by competitors' innovations also are increased simply because of the increased volume of innovative activity worldwide.

TYPOLOGIES OF TECHNOLOGICAL INNOVATION

Freeman has proposed four categories of technological innovation:

1. *Incremental innovations* are those concerned only with improvements in existing products, processes, organizations, and production systems. They are closely linked to actual or potential market demand or experience in use and are driven by the user-producer and learning-by-doing relationships of producers and users. Such innovations follow a well-defined or relatively predictable techno-economic trajectory, and while they may not be dramatic individually, they have a large cumulative impact. Moreover, they often are essential to realizing the potential payoff from radical innovations.

2. *Radical innovations* are those that produce discontinuities in the techno-economic trajectory. They do not arise from incremental improvement of an existing product, process, or system. Indeed, one of their most important impacts is that they change the parameters for cost-effective incremental innovation.

3. *New technological systems* are constellations of innovations that are closely interrelated both technically and economically. Clusters of innovations form "natural trajectories" that are gradually consolidated into a system. Thus over time they become increasingly incremental as the interdependencies deepen and are assimilated into the economic, social, and educational structure.

4. *Changes of technoeconomic paradigm* correspond most closely to the

"creative gales of destruction" in Schumpeter's theory of economic growth under capitalism.[8] Such innovation is accompanied by several clusters of radical and incremental innovations and embodies several different technological systems. But most important are the pervasive effects throughout the entire economy, which include the organizational and social changes and the widespread acceptance of technical and management practices that are necessary to fully realize the impacts of the technical changes.

Most technological innovation is incremental in nature. Here the intimate interaction between users and producers of technology is the key to success in the market. Moreover, it is the cumulative effect of many apparently minor incremental innovations that is primarily responsible for steady growth in productivity and for the expansion of both the size and technological scope of markets. Small, incremental innovations are even more important to economic success in developing countries than in developed countries. The conditions for such incremental innovation are optimized when the technology transfer from vendors is deliberately planned as a learning vehicle for the entire work force of the recipient firm. This will require a negotiating strategy to ensure that engineers and technicians in the recipient firm are involved in all the activities of the suppliers in order to promote the transfer of not only specific know-how but also related generic and systemic knowledge of the relevant technologies. In this way, the firm's own personnel can begin to contribute added value at the earliest possible moment to the information transferred.[9]

The difficulty with such categories of innovation is that they overlap and are complementary so that, for example, radical innovations and, even more, technological paradigm shifts can realize their economic impact only through intensified incremental innovation efforts made after they first appear. Thus the role of radical innovations is largely to open up opportunities for new kinds of incremental innovations by increasing the reward/cost ratio of such innovations, much as the discovery of a new vein of ore in mining expands the economic rewards possible from prospecting that particular lode. But such novel efforts also are open to greater competition since the new paradigm tends to erode the competitive advantage derived from cumulative experience with older paradigms and thereby narrows the window of opportunity for competitive success in incremental innovation. This is an important consideration for developing countries because it implies that the work force must experience continual cumulative learning, both from experience and from formal training, in order to remain competitive in a world market where intense, continual, incremental improvement is increasingly essential to sustained competitiveness.

Similarly, new technological systems generate demand for all kinds of synergistic collateral innovations, some of which may be radical innovations in new fields. If the range of such collateral innovations becomes broad enough and affects enough sectors of economic activity, the whole can grow into a new technoeconomic paradigm, fundamentally altering the structure of relative factor costs and demanding far-reaching social and institutional innovations.

The transition from a single radical innovation to a new technoeconomic paradigm through the evolution of technological systems is seldom anticipated at the beginning of the process; it only becomes apparent gradually over time. Thus information technology and microelectronics, taken together, are a prime example of a new technoeconomic paradigm. Yet it all began with the invention of the transistor in 1947 and the earliest electronic computers that preceded the first use of transistors. Both the computer and the transistor were seen as radical innovations, but initially both were perceived as merely radically improved substitutes for existing technologies—the transistor for the vacuum tube and the computer for the mechanical calculator or tabulating machine. The new technoeconomic paradigm was recognized only gradually as numerous other innovations in materials, solid-state devices, mathematical programming and higher-level computer languages, information theory, and signal processing, among others, emerged at first more or less separately and then merged and became much more widely diffused. Sometimes, a particular radical innovation was recognized in retrospect to have been a precipitating event. The integrated circuit and the microprocessor were such events in the case of the information technology paradigm; the first higher-level computer language was as well. But such events appear to have been precipitory only in light of the availability of many other ancillary technologies.

A new technoeconomic paradigm can be recognized by three features:

1. Its clearly perceived low and rapidly falling cost.
2. Its apparently almost unlimited supply, available for long periods.
3. Its clear potential for use or incorporation in many products and processes throughout the economy, either directly or through related organizational and technical innovations that reduce the cost and enhance the quality of capital, labor, and material to virtually all sociotechnical systems.

Two examples are the mass-production paradigm of the early twentieth century (Fordism) and the "information society" paradigm that the world is still in the midst of (or perhaps only on the threshold of) today. Time has shown, however, that often decades pass before the enhanced productivity potential inherent in paradigm shifts is realized.[10]

One of the aspects of technology transfer that has not been fully sorted out is the relationship between the criteria for successful innovation and the types of innovation. In particular, it is not entirely clear which features are most important for success in which types of innovation and how these features and types interact.

TECHNOLOGY TRAJECTORIES AND LIFE CYCLES

Another important concept is that of the technology trajectory or technology life cycle. Each type of innovation goes through such a life history, and the higher categories of innovation are in some sense the result of the superposition of life cycles associated with components from the lower categories of innovation.[11] The lower categories usually go through their life cycles more rapidly than the higher categories, and the highest category, the technoeconomic paradigm, is often hypothesized to be with the Kondratiev long waves of economic speculation.[12]

Technology historian Thomas P. Hughes distinguishes between two types of radical innovation: those that are radical in a technological sense but fit rather readily into existing institutional structures, and those that can only be realized on a significant scale after a substantial restructuring of institutions or social innovation accompanying technical innovation.[13] Put more crudely, one type of innovation favors existing power structures; the other tends to disrupt them.

A second subcategorization that applies mainly to incremental and radical innovations distinguishes between innovations for which a single patent or group of patents or trade secrets held by a single owner or inventor provides a relatively unassailable proprietary position ("simple" technologies), and innovations whose practical realization depends on a series of interdependent patents or trade secrets not controlled by any single company or organizational entity, thereby requiring extensive cross-licensing or alliances ("complex" technologies).[14] Specific chemical compounds and pharmaceuticals are examples of simple technologies, while most electronic innovations are examples of complex technologies.

Why should radical innovation and the emergence of new technoeconomic paradigms be relevant to problems of development, despite the fact that these types of innovation are likely to be created only by industrialized countries that possess the highly capable scientific and technological infrastructures necessary for such creation? The answer is that both of these types of innovation, the radical and the systemic, can generate many new niches for incremental innovation that do not necessarily require the advanced knowledge and experience needed to create and manage the technological system as a whole. Developing countries with minimum levels of basic education in their work force and little industrial experience may be able to fill these niches quite successfully, often at lower cost than developed countries. This already has been dramatically demonstrated by the success of the Asian "tigers" in numerous niches opened up by the revolution in information technology (see "Information Technology for Development" by John S. Mayo in this volume). Similar niches may be opening in the field of energy efficiency and certain kinds of decentralized generating systems (see "Technology Innovations in Energy" by Richard E. Balzhiser). This development may stem from the so-called Leontief paradox, which asserts that the early stages of a radical innovation are often labor-intensive rather than capital-intensive.[15] Although Leontief's theory was originally advanced to explain the trade advantage of *developed* countries in labor-intensive, high-technology trade, his argument can be modified to suggest that developing countries also can develop a competitive cost advantage when they use their own well-trained engineers and technicians to implement incremental innovations at a much lower cost than if such innovations were implemented by the more expensive engineers and technicians in the developed countries who can generate more value relative to their salaries in more sophisticated fields. Examples are disk drives for personal computers and certain types of software developments that have migrated to the newly industrialized countries where they incur much lower engineering man-hour costs than in the developed countries.

In this kind of competition, timing is critical. Advantage accrues to well-prepared developing countries at a phase in the technology life cycle when demand is expanding rapidly (including the demand for ancillary and supporting technologies, which are the essential components of an emerging technological system), rather than at the earlier critical stage of innovation and discovery that produced the new technology in the first place.

Another kind of opportunity for developing countries which is more speculative arises from the "lock-in" effect that often develops in rapidly expanding technological trajectories.[16] At a certain stage the originators of the cycle may have accumulated large sunk costs in process-specific capital and process- and product-specific training that they are hesitant to write off before the costs are fully offset by the accumulated volume of sales. At this point, there is room for a new entrant with a different approach provided the market is still growing fast enough. But the opportunities for new entrants are usually small in fields where a single player holds a single patent or group of related patents. Where no one player has an impregnable position without access to intellectual property held by others, the opportunities for new players are greater.

In all of the niche-type opportunities just discussed the aspiring entrant must have a fairly thorough generic understanding of the technological system in which a potential new niche may lie. This is one reason why imitation can be said to be the first step toward innovation—but only to the extent that it provides a real window into an entire technological system.

MODELS OF INNOVATION AND TECHNOLOGY TRANSFER WITHIN FIRMS

Since World War II until recently, much of the thinking about technology policy and strategy in the United States was in terms of a linear-sequential type model of innovation (see Figure 1), in which innovation originates in a scientific discovery, which then leads to applied research, followed by development, design, manufacturing, and marketing in an orderly, unidirectional sequence. In this "supply-side" model of technology development, the economic rewards derivable from new technology are limited primarily by its supply. This model is not an unreasonable description of what happened, for example, in the case of nuclear fission, the laser, and the discoveries in molecular biology that led to the emerg-

FIGURE 1 Linear-sequential model of innovation. SOURCE: Stephen J. Kline and Nathan Rosenberg, "An Overview of Innovation," in *The Positive Sum Strategy: Harnessing Technology for Economic Growth*, ed. Ralph Landau and Nathan Rosenberg (Washington, D.C.: National Academy Press, 1986), 286.

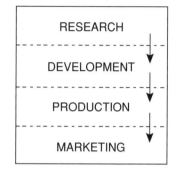

ing biotechnology industry. And as Mayo points out, it represents a fairly good model for technological innovation in telecommunications as it has occurred up until quite recently. But such a linear sequence is occurring less and less frequently as a growing supply of technology accompanies the worldwide growth in the volume of R&D and in the population of scientists and engineers. And when the linear sequence does occur, it is, at the beginning at least, unanticipated. But more important, the model overlooks the two-way or iterative interaction that occurs between the successive stages of technology development, which is significant even in the case of radical innovations that otherwise conform to the linear model more closely than the other types of innovations listed earlier.

A more realistic representation is provided by the so-called "chain-linked" model (see Figure 2). In this model, the early stages in the chain have to be revisited frequently in light of new questions raised when insights are developed only after the downstream phases have been undertaken, including many issues that do not become apparent until early versions of a product or process already have entered the marketplace.

Ken-ichi Imai, one of the leading philosophers of the Japanese theory of innovation, has presented the same general idea in a different way (Figure 3). Imai distinguishes between three strategies of innovation. Type A essentially corresponds to the old linear model in which each phase is completed and then thrown "over the transom" to the next phase with a largely different set of actors taking over. Type B probably corresponds most closely to the current average U.S. practice in which there is substantial overlap between successive phases. Type C represents the predominant Japanese practice in which all three phases overlap much more extensively. Manufacturing and marketing considerations enter into the planning of upstream activities much earlier, and substantial research continues even after first introduction to the market, guided by feedback from customer experience. The best practice of the most successful U.S. companies also has become steadily closer to type C than to type B.

Comparison of the Japanese and American strategies (at least until recently) for the engineering development of a new automobile model (Figure 4) reveals that the Japanese new model development requires about half the number of engineering man-hours needed for the corresponding U.S. development. Furthermore, for Japanese manufacturers the basic features of the overall product and process design are "frozen" much earlier in the cycle. The difference in the Japanese and U.S. patterns is believed to stem in part from the much better job the Japanese do in documenting and codifying their previous model development experiences so that much more experience is transferred from one model development to the next. Similar comparisons have been constructed by Xerox for comparing the Japanese and U.S. development cycles for new photocopier mod-

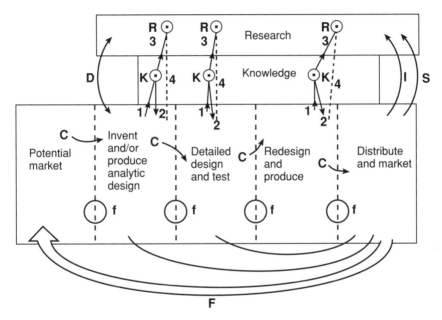

FIGURE 2 Chain-linked model of innovation. Symbols on arrows: C = central-chain-of-innovation; f = feedback loops; F = particularly important feedback.

K-R: Links through knowledge to research and return paths. If problem solved at node K, link 3 to R not activated. Return from research (link 4) is problematic—therefore dashed line.

D: Direct link to and from research from problems in invention and design.

I: Support of scientific research by instruments, machines, tools, and procedures of technology.

S: Support of research in sciences underlying product area to gain information directly and by monitoring outside work. The information obtained may apply anywhere along the chain.

SOURCE: Stephen J. Kline and Nathan Rosenberg, "An Overview of Innovation," in *The Positive Sum Strategy: Harnessing Technology for Economic Growth*, ed. Ralph Landau and Nathan Rosenberg (Washington, D.C.: National Academy Press, 1986), 289.

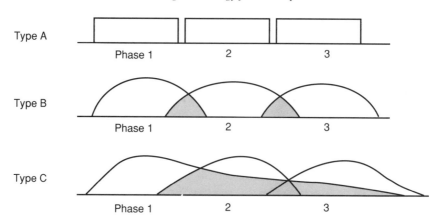

FIGURE 3   Japan's national system of innovation. Sequential (A) vs. overlapping (B and C) phases of development. SOURCE: Ken-ichi Imai, "Japan's National System of Innovation," paper prepared for the NISTEP Conference, Shimoda Tokyo Hotel, Tokyo, February 2-4, 1990.

els. Recent evidence indicates, at least for copiers and automobiles, that the gap in practice has narrowed in recent years and that in a few instances product design for manufacturability early in the cycle has gone further in the United States than in Japan (for example, the Chrysler Neon).[17]

These three representations of innovation all underline the importance of a fully integrated innovation process, which also is consistent with the criteria for success listed earlier. Such integration includes overlapping participation by the actual personnel in the various phases. Thus marketing and manufacturing people might participate in early R&D, and some R&D people might interact with the final customers in the market to obtain feedback for the ongoing improvement *(kaizen)* process that is fundamental to the Japanese system.

The separation between the various stages of the production system generally tends to be a more difficult problem in developing countries than in industrialized countries, in part because of the education gap in developing countries between the engineers and the rest of the work force and often because of a cultural tradition in which "hands-on" work has lower social status than intellectual work, making the integration of the stages of production more problematic. This may be partially counteracted in countries that have a strong artisan and craft tradition that has not yet been extinguished by the spread of the older mass-production paradigm of the West. It also is less of a problem in the countries, such as the East Asian "tigers," that have a strong indigenous tradition of universal, high-quality education at the elementary level. The difficulty can be overcome as well by an emphasis on high-quality, on-the-job training that supports some generic training in parallel with highly job-specific know-how, including inputs from sources other than the exporter or vendor of the technology.[18]

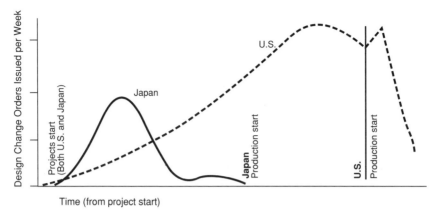

FIGURE 4   Rate of issuance of design changes—patterns of U.S. and Japanese auto manufacturers. Reprinted, with permission, from L. P. Sullivan, "QFD: The Beginning, End, and Problem in Between," in *Quality Function Deployment, A Collection of Presentations and Case Studies* (Dearborn, Mich.: American Supplier Institute, 1987). © 1987 by American Supplier Institute, Inc., Allen Park, Michigan (USA).

It is important to emphasize that the present theory of innovation is very incomplete and impressionistic. Despite a growing volume of detailed case histories, it has proven difficult to develop robust generalizations, and especially to decide the degree to which they apply across many different types of innovation (such as those proposed by Freeman and described earlier) and many different industrial sectors. The generalizations about tight coupling between successive stages in the innovation process and about the importance of awareness of changes in technology and market conditions in the competitive environment are robust when stated at that level of generality, but what this coupling and this awareness actually entail in terms of the actions of managers and workers and technical experts in particular circumstances is far from self-evident. In any given case history, it is often difficult to distinguish between what is idiosyncratic to that particular case and what can be generalized to other similar cases or even what factors are most salient in defining what is meant by "similarity." This is really a part of the unfinished business of innovation research implied in the subtitle of this paper, "Linking Knowledge to Action."

NOTES

1. Nit Chantramonklasri, "Managing Technology Transfer for Acquiring Technological Capabilities in the Context of Developing Countries," draft paper, 1991. Also see by the same author, "The Development of Technological and Managerial Capability in Developing Countries," in *Technology Transfer in the Developing Countries* (New York: Macmillan, 1990).

2. Richard R. Nelson, ed., *National Innovation Systems: A Comparative Analysis* (New York: Oxford University Press, 1993), 4, 6.

3. Chantramonklasri, "Managing Technology Transfer," and "The Development of Technological and Managerial Capability."

4. Robert A. Charpie, chair, Panel on Invention and Innovation, *Technological Innovation: Its Environment and Management* (Washington, D.C.: U.S. Department of Commerce, January 1961).

5. Christopher Freeman, "Networks of Innovators: A Synthesis of Research Issues," *Research Policy* 20, no. 6 (1991), originally presented as a paper at the International Workshop on Networks of Innovators, Montreal, May 1990. Also republished as Chapter 5 in Christopher Freeman, *The Economics of Hope: Essays on Technical Change, Economic Growth and the Environment* (London and New York: Pinter, 1992). For a more recent summary discussion of the history of research on innovation and economic growth, see Richard R. Nelson, "An Agenda for Formal Growth Theory," Working Paper WP-94-85, International Institute for Applied Systems Analysis (IIASA), Laxenburg, Austria, September 1994 (a contribution to IIASA's research program on Systems Analysis of Technological and Economic Dynamics).

6. Chantramonklasri, "Managing Technology Transfer."

7. Eric von Hippel, "Cooperation between Rivals: Informal Know-How Trading," *Research Policy* 16 (1987): 291-302.

8. For a discussion of this and other aspects of Schumpeter's theories of innovation, see Freeman, *Economics of Hope,* chap. 5.

9. Chantramonklasri, "Managing Technology Transfer."

10. Paul A. David, "Computer and Dynamo: The Modern Productivity Paradox in a Not-Too-Distant Mirror," in *Technology and Productivity* (Paris: Organization for Economic Cooperation and Development, 1991).

11. T. Vasko, R. Ayres, and L. Fontvielle, eds., *Life Cycles and Long Waves,* vol. 340 of *Lecture Notes in Economics and Mathematical Systems Series,* ed. M. Beckmann and W. Krelle (New York: Springer-Verlag, 1990). See especially chapter by Harvey Brooks, as well as summary observations by Brooks.

12. Carlotta Perez, "Micro-electronics, Long Waves and World Structural Change," *World Development* 13 (1985): 441-463.

13. S. B. Lundstedt and E. W. Colglazier, Jr., eds., *Managing Innovation* (New York: Pergamon Press, 1981). See chapter by Thomas P. Hughes.

14. Don E. Kash and Robert W. Rycroft, "Two Streams of Technological Innovation: Implications for Policy," draft paper, January 1992.

15. W. Leontief, "Domestic Production and Foreign Trade: The American Capital Position Examined," *Proceedings of the American Philosophical Society* (September 1953).

16. Paul A. David, "Clio and the Economics of QWERTY," *American Economic Review* 75(2): 332-337; and Brian Arthur, "Competing Technologies: An Overview," in *Technical Change and Economic Theory,* ed. G. Dosi et al. (London: Pinter, 1988), chap. 26.

17. *Economist*, October 15, 1994.

18. Chantramonklasri, "Managing Technology Transfer."

Technological Trends and Applications in Biotechnology

RITA COLWELL
President, University of Maryland Biotechnology Institute

Webster's defines biotechnology as "applied biological science."[1] The U.S. government, however, employs a more comprehensive definition: both the old and new biotechnologies comprise "any technique that uses living organisms (or parts of organisms) to make or modify products, to improve plants or animals, or to develop microorganisms for specific uses."[2] The "new" biotechnology has been defined by the U.S. government as "the industrial use of rDNA, cell fusion, and novel bioprocessing techniques."[3] This being said, the definition that in the long run may be the most descriptive relative to the world economy was produced by Vivian Moses and corporate biotechnology pioneer Ronald Cape: "making money with biology."[4]

Biotechnology already has been employed successfully to manufacture new medicines, improve agricultural production, and produce drugs from metabolites of marine organisms, and it shows great promise in such other areas as remediating environmental pollution. But its most rudimentary applications are in fermentation—that is, the use of microorganisms such as molds and bacteria to produce food products. This application is as old as the history of human civilization. Fermentation technology originated in ancient China, where foods were fermented by molds, and in Egypt, where beer brewing and bread-making were combined enterprises.[5] Bread, cheese, yogurt, vinegar, soy sauce, bean curd, beer, and wine are a few examples of the modern products of fermentation.

The unique characteristics of microorganisms have only begun to be exploited to improve life on this planet, taking into account, of course, the role of microorganisms in the cycling of nutrients and in global climatic processes. But the new methods and technologies are only emerging from old ones. For example, by the end of the eighteenth century, farmers had learned to rotate crops in order to plant crops that restored nutrients to nutrient-poor soil. And even before the science of genetics was understood, new varieties of crops and animals were being bred by selection for desired qualities.

Some milestones in the history of science indicate the source of this new technology. In the field of medicine, Edward Jenner, who in the last decade of the eighteenth century observed that milkmaids did not succumb to smallpox, began inoculation with cowpox, or *Vaccinia* virus, to prevent smallpox infection. About the same time, Louis Pasteur, best known for his work that led to the process of pasteurization and the identification of microorganisms as causative agents of disease, studied fermentation in wine and wrote an important book on winemaking. And around the turn of the century, German bacteriologist and physician Robert Koch identified microorganisms as the causes of anthrax and tuberculosis.

The first industrial use of a pure culture of a bacterium was accomplished by Chaim Weizmann in 1917 when he developed the fermentation of cornstarch by the bacterium *Clostridium acetobutylicum*, thereby producing acetone for explosives manufacture. Gregor Mendel, an Austrian monk whose studies on the pea plant elucidated inheritance of traits via hereditary factors, conducted seminal work in genetics in 1865. Although Mendel's work was ignored until 1900, his findings, once rediscovered, fit well with what was by then known about chromosomal activity during cell division, or mitosis. The first half of the twentieth century was an exciting time, with major gains in knowledge of genetic inheritance. Thomas Hunt Morgan of Columbia University, working with the fruit fly, *Drosophila melanogaster*, showed that genes, or the units of heredity, were the constructs of chromosomes. His student A. H. Sturtevant, who later joined him when he moved to the California Institute of Technology, made a number of breakthrough discoveries showing genes were linked, comprising chromosomes.[6] He thus began the science of genetic mapping, a technique essential to the new genetics.

In the 1930s and 1940s, genetics research was moving inexorably in the direction of the upcoming explosion of knowledge at the molecular level. Researchers such as Barbara McClintock[7] and Marcus Rhoades[8] studied linkage and mutable characteristics in maize (corn), providing a view of genes as more mutable and variable than the simple Mendelian genetics allowed. Meanwhile, research on what comprised genetic material moved forward rapidly. In 1928, Frederick Griffith had found that a "transforming principle" was able to alter traits in the bacterium *Streptococcus pneumoniae*.[9] By 1944, Oswald Avery, Colin MacLeod, and Maclyn McCarty of the Rockefeller Institute had identified the "transforming factor" as deoxyribonucleic acid, or DNA.[10] From that moment, scientists in many laboratories labored to determine the chemical structure of the DNA molecule. Finally, in 1953, James Watson and Francis Crick's short paper described the breakthrough for which everyone was waiting.[11]

GENETIC ENGINEERING: A NEW WORLD

Twenty years after Watson and Crick's paper, the first stones were laid in the path to commercial genetic engineering. Stanley Cohen of Stanford University, Herbert Boyer of the University of California (San Francisco) Medical School, and their teams succeeded in cloning a gene into a bacterial plasmid—the first recombinant DNA (rDNA)—and in 1980 they received a patent for this technique.[12] Also in 1980, the U.S. Supreme Court ruled in *Diamond v. Chakrabarty* that microorganisms could be patented, opening a new commercial avenue for genetic engineering.[13]

The first U.S. biotechnology company, Genentech, was founded in 1976. By 1994, it had been joined by more than 1,300 other companies in the United States alone (Figure 1).[14] The years 1981-1987 were watershed ones for U.S. biotechnology: an average of 90 companies were formed annually, for a total of 631 companies established during this period.[15] In 1981, the first U.S.-approved biotechnology product reached consumers: a monoclonal antibody-based diagnostic test kit. The following year, the first recombinant DNA pharmaceutical, Genentech's Humulin (recombinant human insulin), was approved for sale in the United States and Great Britain. Humulin's 1993 sales were $560 million.[16] The same year, the first recombinant animal vaccine for colibacillosis was approved in Europe.

Although most biotechnology companies are still not consistently profitable, more and more products are entering the market.[17] In 1993, Amgen's Neupogen,

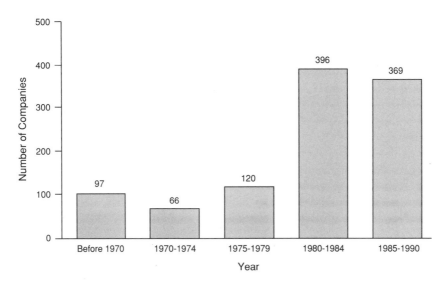

FIGURE 1 Evolution of the U.S. biotechnology industry.

TABLE 1 Top Ten Biotechnology Drugs, 1993

Product	Developer/Marketer	1993 Net Sales (millions)
Neupogen	Amgen/Amgen	$ 719
Epogen	Amgen/Amgen	587
Intron A	Biogen/Schering-Plough	572
Humulin	Genentech/Eli Lilly	560
Procrit	Amgen/Ortho Biotech	500
Engerix-B	Genentech/SKB	480
RecombiNAK HB	Chiron/Merck	245
Activase	Genentech/Genentech	236
Protropin	Genentech/Genentech	217
Roferon-A	Genentech/Hoffman-La Roche	172
Total sales		$4,288

SOURCE: Ernst and Young LLP.

human granulocyte colony-stimulating factor, was the best-selling U.S. biotechnology drug, netting $719 million (Table 1).[18] In 1994, at least four new drugs were approved in the United States, along with a recombinant "housekeeping" enzyme and diagnostics. One new product in another area was Calgene's Flavr Savr tomato, engineered for better taste and shipping tolerance through the addition of a "backwards" gene that induces the tomato to produce only small amounts of the ripening enzyme, polygalacturonase. Thus the tomato can be picked before ripening and left to ripen slowly without adding artificial ripeners.[19] Although it did not need approval at the time of its development, the recombinant tomato underwent review and was approved in 1995 by the U.S. Food and Drug Administration (FDA).

In 1994, U.S. biotechnology companies had a market value of $41 billion, R&D expenditures of $7 billion, and 103,000 employees—this in an industry that did not exist 20 years ago. By comparison, the U.S. pharmaceutical industry, which has invested heavily in biotechnology, had R&D expenditures of $13.8 billion in 1994. Poor economic markets and policy questions in the United States held down the number of companies formed in 1994, but instead of being in a downturn, the U.S. biotechnology industry may be maturing, to eventually take on a new role in the global economy. And instead of being aggressively entrepreneurial, with the intention of becoming the next Merck, the newly emerging companies may well serve in the future as a reservoir of corporate research for large pharmaceutical firms, which, in turn, will develop and market the output.

Today, however, because most of the U.S. biotechnology industry is centered on health care products and many of the companies were started on the basis of licensing agreements or research from the university community, the decrease in corporate start-ups, as well as financing, is causing a basic change in the structure of the industry. Smaller companies are merging; large companies, such as the major pharmaceutical companies, are acquiring smaller biotechnology ventures; and, because there is little money available in the investment market for corporate growth, companies are looking to strategic alliances, both in the United States and abroad, to shore up finances and financial opportunities. This development may prove beneficial for Asian pharmaceutical or biotechnology companies looking for products in return for allowing access to the Asian market. But many of the developing countries, lacking homegrown pharmaceutical giants, will have to look elsewhere for role models for their own fledgling biotechnology industries.

The United States is not the sole benefactor of biotechnology growth. In 1993, 386 biotechnology companies were located in Europe, most in Great Britain, Germany, Belgium, and the Netherlands (Figure 2).[20] From 1986 to 1992, about $657 million was pumped by venture capitalists into the European biotechnology industry. The major biotechnology players in Western Europe are Belgium, Denmark, France (whose 1991 market for biotechnology products was $115 million, $29 million of which were imports), Germany, Italy (with a 1995 biotechnology market estimated at $1.5 billion), the Netherlands (whose 1991 biotechnology product and process sales equaled $220 million), Sweden, and the United Kingdom. In 1993, Canada had 310 biotechnology companies, with revenues of $1.67 billion and 61 percent of total sales from exports.[21] Ten percent of Canada's biotechnology exports go to Japan, while an additional 10 percent go to China, India, South America, and the Caribbean. A handful of companies are

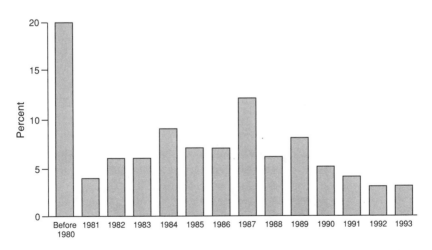

FIGURE 2   Evolution of European biotechnology industry (percent of companies founded in each year). SOURCE: Ernst and Young LLP.

scattered in South and Central America (mainly in Brazil and Mexico) and Asia (excluding Japan). Approximately 200 biotechnology companies are located in Australia and an additional 40 in New Zealand. Japan's biotechnology industry differs from the entrepreneurial industry in the United States, Canada, Europe, and Australia in that much of Japan's biotechnology R&D is carried out by universities and research institutes or in cooperation with its large pharmaceutical firms, food corporations, brewing companies, or electronics giants. The R&D outlays of Japan's top ten pharmaceutical companies are only one-fifth of similar outlays by U.S. companies.[22]

In most developing nations, there is little in the way of commercial biotechnology, but governments and researchers acknowledge the importance of the field, and government and nongovernmental organization support have led to establishment of biotechnology-related centers. For example, the International Center for Genetic Engineering and Biotechnology (ICGEB), initiated by the United Nations Industrial Development Organization (UNIDO) but now supported by Italy and India, has two laboratories: one in Trieste, Italy, and the other in New Delhi. Research groups from 32 member countries are affiliated with ICGEB.[23]

The M. S. Swaminathan Research Foundation in Madras is a leader in the promotion of biotechnology at the village level in India. Other prominent biotechnology research institutes in India are New Delhi's Energy Research Institute and a national institute of cellular and molecular biology in Hyderabad. Hindustan Lever, a subsidiary of Lever Brothers, has a large corporate biotechnology division in India (Kamaljit Bawa, University of Massachusetts, personal communication, October 28, 1994).

In the Far East, the Hong Kong Institute of Biotechnology, under the auspices of the Chinese University of Hong Kong, was established with the help of overseas Chinese scientists.[24] Hong Kong also has a Biotechnology Research Institute. Of the many biotechnology-related research departments and institutes in the People's Republic of China, one of the oldest and best known is the Shanghai Institute of Biochemistry.[25] The International Vaccine Institute being established in South Korea is receiving financial assistance from the United Nations Development Fund and the Japanese government.[26] Thailand's National Centre for Genetic Engineering and Biotechnology, which has a marine biotechnology laboratory, was begun with support from the U.S. Agency for International Development (USAID). Aquaculture is a major theme of other biotechnology research centers and university departments in Thailand.

Worldwide, many national and international organizations maintain laboratories that carry out research in biotechnology, mostly related to agriculture. One example is the International Rice Research Institute in the Philippines. Another is the Biotechnology Centre for Animal and Plant Health, established by the European Union, in partnership with the Queen's University of Northern Ireland in Belfast, which focuses primarily on disease control.[27]

In Africa, the French molecular genetics researcher Daniel Cohen and his organization, Association Ifriqya, are planning to establish the Institute for Genome Research for Developing Countries (IGRDC) in Hammamet, Tunisia, in 1996.[28] The Agricultural Genetic Engineering Research Institute (AGERI) in Cairo, Egypt, is cooperating with Michigan State University's Agricultural Biotechnology for Sustainable Productivity (ABSP) project, supported by USAID.[29] The African network of Microbiological Resources Centres (MIRCENs), although not a research group per se, is organized to support research projects in soil microbiology, biotechnology, natural resources management, vegetable production and protection, and food and nutritional technology at research organizations and universities throughout Africa south of the Sahara.[30] The International Laboratory for Research on Animal Diseases (ILRAD) is located in Nairobi, Kenya.

MARKET SEGMENTS AND RESEARCH AREAS

In the United States, and to a lesser degree in Canada and Europe, the bulk of the biotechnology industry is in the biomedical field: therapeutics and diagnostics make up 68 percent of the U.S. industry, 43.7 percent of the Canadian industry, and approximately 43 percent of the European industry. From 1993 to 1994, therapeutic product sales in the United States increased 24 percent, for a total of nearly $20 billion.

Agricultural biotechnology also represents a growing segment of the industry: 8 percent in the United States, 20 percent in Europe, and 28 percent in Canada. The U.S. agricultural biotechnology market increased its sales by 158 percent in 1993, with aquaculture the most rapidly growing sector.[31]

The chemical, environmental, and services segment, which makes up only 9 percent of the U.S. biotechnology industry, increased its sales by 81 percent in 1993, totaling $70 billion.[32] This segment comprises 10 percent of the Canadian industry.

Biomedical

Medical biotechnology mainly includes recombinant drugs and enzyme-mediated diagnostic kits, but the rational design of drugs, where a drug is modeled to fit a particular molecule, yielding a limited response that can result in control of the disease process, has become a significant part of this field. By learning more about the basic biochemistry of normal and abnormal cellular function, scientists eventually will produce drugs that will prevent the abnormal growth of cancer cells, or will permit detection of the abnormalities in the DNA that signal the onset of cancerous changes, thereby preventing cancer from occurring. Another intent is to circumvent the immune response to one's own tissue that occurs in such autoimmune diseases as multiple sclerosis and lupus erythematosus. The hope also is to use small molecules to combat degenerative neurological diseases or to induce neurological cell regrowth in such conditions as Alzheimer's disease, amyotrophic lateral sclerosis, head and spinal injury, and cerebrovascular accident or stroke.[33] Some of the already successful recombinant drugs include recombinant human insulin, growth hormone, interferons, tissue plasminogen activator, erythropoietin, and other blood cell-stimulating factors. Thus biotechnology and pharmaceutical companies legitimately have high hopes for the economic and medical potential of the next generation of drugs.

Among the most successful of the antibody-based diagnostics are pregnancy test kits, which now are so simple that in the United States they can be purchased over the counter and used at home. Human immunodeficiency virus (HIV) test kits are being sold worldwide and are manufactured in many parts of the world. The U.S. market for monoclonal antibodies, the majority of which are used in such test kits, was estimated at $1.2 billion this year and to be nearly $4 billion by the turn of the century. As test kits become both more accurate and easier to use, test kit manufacturers foresee wide applicability, even in rural settings, by technicians with minimal training. Some companies, for example, have sent personnel to China and South America to train technicians in the proper use of their test kits.

Monoclonal antibodies were expected to become major tools for the treatment of a variety of diseases, but recent problems with monoclonal antibody-based septic shock treatments caused several companies such as Xoma, Centocor, Chiron, and Synergen to abandon drugs in clinical trials.[34]

Recombinant vaccines are expected to make a major contribution to the health of the world's population. Recombinant hepatitis B vaccine already is used worldwide. Although an HIV vaccine would have enormous use, especially in those countries where HIV is widespread, little success has been achieved and not much is on the horizon, at least at present. Research on HIV vaccines that would be beneficial to those people outside the developed nations, who suffer from a different strain of HIV than that found in the United States and western Europe, is not being pursued aggressively.[35] In contrast, vaccines against malaria, respiratory syncytial virus (RSV), rotavirus (which causes severe, life-threatening diarrhea in children), *Streptococcus pneumoniae* (which causes bacterial pneumonia), and cholera, are being pursued actively and will have an immediate impact on global health.[36] Although new vaccines and vaccine combinations could improve the health of many children worldwide,[37] a recent study showed that the world vaccine market stands at a mere $3 billion,[38] a relatively insignificant value when compared to the $1.2 billion world sales of just one new biotechnology drug, recombinant human erythropoietin (Amgen's Epogen).[39]

Drug delivery systems are an important segment of the biomedical component of the biotechnology industry.[40] New methods of administering vaccines—by injection, intra-nasally by spray, time-release methods, and others still under development—and even drugs, could revolutionize health care in developing nations and in poor or rural communities in developed countries.[41]

Other, much smaller and more specialized medical biotechnology markets include treatment regimens, such as gene therapy. In gene therapy, which currently is employed for research purposes only, a normal gene is put into abnormal cells using a carrier such as a virus. In "cellular therapy" a patient's cells are treated. An example of this is autologous bone marrow transplants, where a patient's bone marrow is removed, cleansed of cancer cells if they are present, grown in tissue culture, then reinjected into the patient—who usually has an advanced cancer—after the patient undergoes therapy to destroy the remaining bone marrow. Because such techniques are prohibitively expensive, they are used sparingly. Gene therapy requires high-technology medical centers and a high level of training for all staff members involved in patient care. Clearly, even in developed nations these treatments are available only to the very wealthy, the very well insured, or enrollees in sponsored clinical trials.

Agriculture

Agricultural biotechnology is expected to become the predominant application of biotechnology in developing countries. In Africa,[42] Asia, Central and South America,[43] and the Middle East,[44] development of transgenic plants, biological pest control, tissue culture techniques for agriculture, microbial products for nutrient cycling, pathogen diagnostics for crops, and genetic mapping of tropical crops are major concerns. In developed nations, the term *value added* is used to denote the economic value of agricultural biotechnology products. Thus agricultural biotechnology in the United States, Canada, Europe, Japan, and Australia aims to produce products, such as fruit, vegetables and grains, whose genetic manipulation will provide new products that will cost more or bring greater profit to commercial entities than the standard hybrid product. Today in the United States the best-known commercial agricultural biotechnology products include Calgene's Flavr Savr tomato; Monsanto's recombinant bovine somatotrophin (BST) or growth hormone, which yields increased milk production by cows; frost-resistant strawberries; and biological pest control, which may include the introduction of genes from *Bacillus thuringiensis* and other bacteria, fungi, or viruses into plants, rendering the plants pest-resistant,[45] and the production of biopesticides via gene isolation and fermentation. Less well known is the production of recombinant rennin, an enzyme used in cheese manufacture, approved by FDA in 1990.[46]

Transgenic plants are those in which foreign genes have been introduced to improve a specific quality or characteristic of the plant. In the case of the Flavr Savr tomato, transportability is improved, an important factor in areas where fruits and vegetables must be transported long distances to market. Since the developed nations—such as the United States—are dependent on Central and South American countries for fruits, especially during winter and early spring, before the harvests in Florida and California, these technologies could increase the marketability of imported crops. Moreover, the introduction of foreign DNA could improve the protein quality of some foods, an important consideration for developing countries not only for human foods but also for animal feeds. Researchers also are working on improving the nutritional qualities of food starches and oils.

Biological pest control, a technique used in Asia for several millennia, has made great strides since the United States began to import *B. thuringiensis* into the United States from China in the late 1970s.[47] *B. thuringiensis* is engineered for inclusion in many plants, including grains, and it also is manufactured by recombinant techniques for use as a spray. Other means of biological pest control include virus resistance incorporated into the plant genome.[48] A virus-resistant squash is being reviewed by the USDA for approval in the United States; China is marketing a virus-resistant tomato; and potatoes resistant to virus are undergoing testing in Mexico. Scientists in Costa Rica are working to introduce virus-resistant genes into the criollo melon.[49] Recently, investigators identified a number of genes within crops themselves that confer disease resistance.[50] Thus it is only a matter of time until such genes are introduced into nonresistant species.

An important field in agricultural biotechnology will be the use of marker or "reporter" genes within transgenic species.[51] These genes are attached to functional genes introduced into plant cells, where their presence will indicate if the functional genes are working. Recently, researchers at the U.S. Department of Agriculture and University of Wisconsin inserted a gene for green fluorescent protein, derived from the jellyfish, *Aequorea victoria*, into orange tree cells.[52] This is a unique melding of agricultural and marine biotechnology and is an early example of more unique genetic introductions to come.

Much of the improvement in crops depends on improved plant tissue culture techniques and techniques for plant micropropagation. In tissue culture, individual cells are separated, genetically modified for desirable traits, and grown on nutrient media. Hormonal growth enhancers, nutrients (some of which are produced by tissue culture), and other additives determine the viability of the cells maintained in culture. In micropropagation, tiny plantlets are grown from cells started in tissue culture, all genetically the same, for distribution to farmers.

Agricultural production can be increased not only by direct manipulation of plants, but also by the addition of naturally occurring or genetically manipulated microorganisms.[53] Some of these organisms can be grown in batch fermenters; others require nurturing on host plants.

Agricultural products do not necessarily result in food products for the consumer market. Better plastics and biodegradable disposable items may be produced from plant extracts or refuse. Plant refuse, such as corn husks and stalks, also can be used to produce alcohols and other fuels such as ethanol. Finally, plants and animals can be genetically engineered to produce drugs and other biologically active molecules. In fact, the entire tobacco program of the USDA is now funded only for research on production of bioactive compounds from transgenic tobacco plants.

Not to be forgotten, plants, like humans, become diseased. Thus it is essential that simple diagnostic tests be developed for early detection of disease.

Marine Biotechnology

Marine biotechnology, which represents a small segment of the biotechnology industry—in the United States, approximately 85 companies or about 7 percent of all biotechnology companies—has applications in medicine, agriculture, materials science, natural products chemistry, and bioremediation. Because of the proximity of most of the world's tropical nations to the oceans, as well as their climates, these nations are particularly well suited to pursue marine biotechnology.

Aquaculture, a branch of marine biotechnology, is closely related to agriculture and is often included under that classification. Worldwide, marine aquaculture produced 14 million metric tons of fish in 1991,[54] with a market value of approximately $28 billion.[55] Demand for seafood worldwide is expected to increase by 70 percent over the next 35 years,[56] but this increase comes at a time when the world's fisheries are overexploited or "commercially extinct."[57] Thus world aquaculture will need to increase production sevenfold by the year 2025 in order to meet the demand. USDA has predicted that biotechnology will aid in the improvement of captive management and reproduction of species, leading to more efficient species that make better use of food supplies and the production of healthier organisms with improved food and nutritional qualities. Furthermore, aquaculture can produce organisms for use as biomedical models in research, reservoirs for bioactive molecule production, and agents useful in bioremediation. Aquaculture is no longer a means of producing luxury foods, such as lobsters; it is a critical solution to the world's fisheries problems.

Algal aquaculture, an ancient art in Asia, produces not only seaweeds, but also food supplements, such as the omega-3 fatty acids and beta carotene, through microalgal cultivation.[58] The polysaccharides of algae are a valuable commodity and a much sought-after natural product.

Animal Husbandry

One of the first approved biotechnology products offered on the market was an rDNA vaccine against colibacillosis. Thus animal husbandry was among the first sectors into which a commercial biotechnology product was introduced. Transgenic animals, such as pigs and cows, can be engineered for traits allowing better survival in marginal habitats, the production of more meat of higher quality, or even the production of recombinant pharmaceutical molecules for the human health care market.

Techniques for *in vitro* fertilization (IVF) were perfected in cattle, allowing breeders to produce multiple embryos from the cows and steers with the best qualities. Cows other than the genetic mothers can then carry the offspring. Biotechnology also can improve the health of these animals with new vaccines and diagnostic methods,[59] leading to increased trade in meat, animal products, and live animals (trade now is often restricted because of fears of spread of disease). For example, USDA, the Yale University School of Medicine, and Virogenetics, Inc. (a Troy, N.Y., biotechnology company specializing in vaccines) have produced a genetically engineered vaccine against Japanese encephalitis in swine, using *Vaccinia* virus and canarypox virus,[60] that is now undergoing field tests. Because the United States is concerned about importation of this disease from Asia, the market for the vaccine may be significant.

Single-Cell Protein

The production of single-cell protein (SCP)—a mass of microorganisms along with their nutrient contents—for both animal feed and human food has at least a 30-year history.[61] Initially, hydrocarbons were used as source material for the nutrients, but higher prices for petroleum rendered SCP non-cost-effective, thereby slowing research on SCP over the past two decades. Bacteria and yeasts have been used to ferment petroleum products, methanol, methane gas, lignocellulose, spent sulfite liquor byproducts of paper mills, molasses, whey, and other industrial fermentation byproducts. Increased efficiency, however, will be necessary before such processes are viable economically.

Environmental Biotechnology

Bioremediation represents a large market force in biotechnology, but its potential has only recently been recognized. For example, U.S. federal laws requiring cleanup of toxic waste sites, surface-mining areas,[62] watersheds, and other polluted sites have caused the U.S. environmental remediation market to expand exponentially.[63] Indeed, in 1993 product sales increased in the chemical, environmental, and services category by 81 percent, to a total of $69.9 billion.[64]

According to a recent National Research Council report, the U.S. bioremediation market will continue to grow rapidly and should reach $500 million a year by the year 2000.[65] A less conservative estimate is that $1.7 trillion will be spent on remediation of hazardous waste sites in the United States within the next 30 years![66] This market does not even include known contamination sites in former Soviet bloc nations. Western nations have offered—but not paid—close to $1 billion for cleanup of these areas,[67] and clearly a sizable percentage of these sites will undergo some bioremediation. In 1993, 10 percent of Canada's biotechnology companies were working in environment-related areas—waste management, biomass, remediation and recycling, and materials reuse.[68] Europe, unfor-

tunately, has as yet too few firms in environmental biotechnology for a statisti-
cally valid evaluation.[69]

Both naturally occurring organisms and genetically modified organisms
(GMOs), especially microorganisms, are used in environmental bioremediation.
Current practice includes altering the environment of the naturally occurring microorganisms to make them work more efficiently; "bioaugmentation," which, in
general, involves the addition of nutrients, most commonly nitrogen and phosphorus; and controlling the oxygen and water contact.[70] Hydrocarbon contamination—
oil spills—is currently remediated using this technology. Other contaminants,
however, are more recalcitrant. Some of the aromatic compounds, polychlorinated
biphenyls (PCBs), and other substances can be removed using genetically engineered microorganisms (GEMs), modified to degrade the target substance or to
function in a particular type of environment.[71] The naturally occurring white rot
fungus, *Phanerochaete chrysosporium,* can degrade PCBs, DDT, cyanide, TNT,
and other toxic soil pollutants.[72] Cellular components, such as enzymes and
biological surfactants, also can be used for environmental cleanup.

The *Exxon Valdez* oil spill cleanup provided a valuable case study in bioremediation. The application of oleophilic fertilizer resulted in enhancement of
biodegradation through enrichment of those microorganisms that degrade oil,[73]
although some questions remain about the efficacy of this technique.[74] Other
fertilizers also were used, along with specific nutrient enhancement and the addition of microorganisms.[75] The conclusion: bioaugmentation clearly was effective, but the addition of microorganisms has yet to be proven of value.

Many biological methods have been proposed for treatment of contaminated
sites. For example, one method familiar to many backyard gardeners, home-
owners, and farmers is composting, where bacteria and fungi decompose organic
material; another is treating polluted soils in the presence of oxygen.[76]
Composting has been used to clean oiled shoreline waste[77] and soil contaminated
with TNT.[78] For some techniques that can be utilized *in situ,* where the contamination occurs, oxygen and nutrients are injected, using specialized equipment.
Monitoring equipment may have to be brought on-site to establish efficacy and to
provide a means of controlling degradation events.

In phytoremediation, heavy metals are removed from contaminated soils by
plants that take up the metals and concentrate them. The plants then can be
burned, both to recycle the metals by producing ores and to produce electricity.[79]
Researchers are now studying how to produce transgenic plants with improved
metal uptake capabilities.

Ex situ remediation is carried out in a bioreactor or filtration system, sometimes in a processing plant or other facility but not necessarily on-site. One
interesting combination of *in situ* and *ex situ* bioremediation is the use of Sea
Sweep, an absorbent composed of a treated material made from wood chips.
After its use in cleaning oil spills, the material is gathered up and degraded by
composting.[80]

Soybean and rice hulls, rice bran, and sugar beet pulp have been found to bind metals and other industrial waste products and may prove useful in the environmental remediation[81] of solid or semisolid waste, liquid waste, sewage, and industrial and agricultural wastes. Many methods exist, including bioreactors and biofiltration.

The biological treatment of raw sewage is widely employed and is an example of how environmental cleanup can result in improved public health. The next step, which municipalities and cities throughout the world face, is what to do with the sludge that remains, as well as solid wastes (garbage). In some communities, sludge is sold for conversion to fertilizer.[82]

Water treatments include not only treatment of wastewater, but also treatment of polluted natural bodies of water. *In situ* treatments consist of the use of microorganisms and localized bioreactors; *ex situ* treatments are carried out in wastewater treatment plants. Microorganisms are being modified for use in wastewater treatment, and new methods—such as immobilization of microorganisms—are being developed to achieve increased microorganism contact with the biomass. Anaerobic wastewater treatment employing methanogenic bacteria that produce methane as a byproduct[83] may be especially useful where an affordable supply of energy, the methane gas, is needed.

The environmental test kit market is growing as kits become smaller and easier to use and portable field testing equipment, such as ion chromatographs, becomes available. Thus environmental monitoring can be carried out, with the use of sensors, on a continuous basis. A town in the Czech Republic, for example, has attached sensors to an illuminated scoreboard that gives continual readings of air pollutants.[84]

Another aspect of environmental biotechnology is improvement of air quality and prevention of both the carbon dioxide buildup and ozone depletion that occur when pollutants are discharged to the atmosphere. The environmental biotechnology company Envirogen is studying organisms for bioremediation of air contaminated with halogenated hydrocarbons.[85]

Ore extraction has caused massive environmental degradation in many parts of the world. For example, Brazil's waterways are polluted with mercury as a result of gold mining. In northern Russia, within the Arctic Circle and near its border with Finland, as much as 2,000 km^2 of forest were destroyed by the sulfurous byproducts of nickel mining.[86] Engineering of microorganisms or the use of naturally occurring microorganisms to remove ores is likely to reduce or eliminate this kind of pollution at mine sites, as well as play a role in the bioremediation of the environmentally impacted mining regions.

Silviculture and the Role of Forests

The world's forests are being destroyed at a frighteningly rapid pace, the swiftest in history. Mature tropical forests, which are estimated to have covered

1.5-1.6 billion ha of the earth's surface, have been slashed in half, and forested land areas continue to decrease.[87] Canada is one of the leaders in biotechnology-based forestry services, earning $25 billion in 1992. Although most of this income is from the paper and pulp industry, bioremediation of effluent and the addition of bacteria to the paper-making process to decrease toxic effluent and improve paper quality are important research goals. Canadian researchers also are working on trees produced via tissue culture to aid in forest restoration. Investigators in Europe, Canada, and the United States have found that invading forest weeds capable of destroying native forest understory or preventing growth of young trees can be controlled by mycoherbicides.[88]

Increased carbon dioxide has a major effect on the world's forests[89] since 90 percent of all carbon contained in terrestrial vegetation is in the forests. Increased atmospheric carbon results in increased growth of temperate and boreal forests. Thus it has been suggested atmospheric carbon be decreased by reducing the use of fossil fuels and increasing the use of biomass-based fuels that release little or no carbon dioxide. Another suggestion is that massive, managed tree plantations be established.[90] Although the latter is not feasible, biomass energy production is a particularly attractive method for developing countries. Studies to determine the effects of increased carbon dioxide are under way, and researchers are studying microorganisms associated with forest trees to devise new methods of altering carbon partitioning.[91] In the United States, the Electric Power Research Institute is analyzing the use of halophytic, or salt-tolerant, plants to sequester carbon dioxide. These plants have added potential as biofuel and to remediate toxic wastewater.[92]

Nonagricultural Marine Biotechnology

The oceans represent the last great frontier for the discovery of new materials, medicines, and foods. Marine biotechnology can be applied in many areas outside those related to food production.[93] Marine natural products are applicable in fields as far ranging as molecular biology and bioremediation, to adhesives and pharmaceuticals. Enzymes isolated from thermophilic *Archaea*, microorganisms originally thought to be bacteria, some of which live in deepsea hydrothermal vents, are essential to molecular geneticists doing DNA sequencing. Agar, an important ingredient in nutrient substrates for growth of microorganisms in culture, and agarose, used to make gels for biochemical genetics and protein studies, are derived from algae.

The marine bacterium *Acinetobacter calcoaceticus RAG-1* emulsifies hydrocarbons. Metal-concentrating marine bacteria also have been identified and may prove useful in marine bioremediation. The strength of adhesives produced by such marine organisms as mussels and barnacles has been recognized, and with the advent of modern biomolecular techniques scientists have been able to study and duplicate some of these materials. Some of the most potent natural toxins known to science are produced by marine organisms. These toxins could be used in research applications, such as studies of the neuromuscular junction, where much of their toxic activity is concentrated. They also could yield potent anti-neoplastic drugs. All these possibilities confirm that monitoring of the marine environment may yield clues about environmental degradation and that studies of marine ecology, including the problem of pollution of shorelines by bacterial pathogens, will provide improved human health.

Energy Production

Although a relatively minor consideration for developed nations at this time, energy production from biological waste products will prove important in the future, at first for developing nations, and later for those countries that no longer can afford to depend on petroleum products.[94] Production of methane from biogas digestors can be carried out on a local or industrial scale. A variety of hexose sugars can be used for the fermentation production of ethanol, but major sources are sugarcane, maize, wood, cassava, sorghum, Jerusalem artichoke, and grains. Waste whey also may be used. Bioconversion processes yield such byproducts as single-cell protein and enzymes for biocatalysis.[95]

Other Areas

On the cutting edge of biotechnology research—currently too small to be even a blip in the marketplace—are biosensors, bioelectronics, biomaterials, and biocomputing (the use of biomolecules in electronic equipment), as well as the development of molecular machines or submicroscopic molecules, some of which are biological in origin, to carry out specific mechanical or energetic functions within the body.[96]

Biosensors have applications in medicine, especially in diagnosis and therapeutics;[97] in process control, where biosensors could be used to determine changes in pH, conductivity, molecular concentration, or other measurable phenomena;[98] in bioremediation, where bioluminescent organisms could function as reporters;[99] and as environmental sensors.[100] For military use, biosensors could be linked to biocouplers to transmit a sensed event, via a biochip, to a computer system.[101] These could be used for environmental monitoring, terrain monitoring, or monitoring of personnel. They also could be used to detect chemical, toxic, and biological warfare (CTBW) agents.

The military could use biomaterials as protective clothing against CTBW agents, or as medical materials such as artificial bone and other tissue, or even as agents of warfare, causing engine malfunctions in enemy vehicles.[102] Nanomachines produced from biological molecules may be used as biosensors, in nano-scale manufacturing processes,[103] or even as methods of drug delivery.[104]

Biocomputing will use biological material and reactions in computer chips.

Bioinformatics—the development of information systems on biology—is a world-wide effort in which all nations, no matter what their developmental stage, can participate.

Safety and Public Acceptance of
Genetically Engineered Products

Monsanto's recombinant bovine somatotrophin, a drug expected to increase the milk production of a herd of cows by approximately 10-20 percent, received FDA approval in 1994, but a campaign against its use began long before BST was approved. Concerns were expressed that people who drank milk produced by BST-treated cows would be affected by the hormone and that these cows would be more likely to develop an infectious mastitis, requiring antibiotic treatment and thereby adding antibiotics to the milk supply. Although it has been concluded by the FDA and others[105] that BST is safe, an economic consequence of treating more than 800,000 of the 9.5 million cows in the United States with the hormone has been an increase in milk production and a decrease in milk prices.[106]

For plants, it is feared that genetically engineered crops will become weeds or will transfer the introduced genes to native crops, which, in turn, would become weeds.[107] Another concern is that the genetically engineered crop itself may become a pest.[108] Other fears are that plants genetically engineered for virus resistance will cause the emergence of new viral pathogens that could affect other crops, that plants genetically engineered to produce toxins may inadvertently cause illness or death in animals feeding on them, and that engineered plants may out-compete wild plant species, altering habitats and affecting other species within those habitats.[109]

To allay fears and guarantee that genetically modified organisms are employed responsibly, field testing should be done safely so that GMOs pose little or no risk;[110] indiscriminate field testing of GMOs should be discouraged. Investigators involved in field testing of GMOs should adhere to UNIDO's "Voluntary Code of Conduct for the Release of Organisms into the Environment," as well as local and national regulations.[111]

Public acceptance of the products of genetic engineering is a major obstacle to be overcome. For example, some chefs in the United States have banded together to boycott the Flavr Savr tomato, and there have been requests that milk from BST-treated cows be labeled as such, or, conversely, that milk from cows not treated with BST be so labeled, although FDA discourages such labeling.[112] Consumers Union opposes the use of BST and the European Union has banned its use.[113] Ironically, most Americans are unaware that some cheeses are manufactured using recombinant rennin.

It is hoped that public education will allay some of these fears.[114] Farmers need factual information as well to assist them in deciding to use GMOs and other products of biotechnology.[115]

BIOTECHNOLOGY'S POTENTIAL IMPACT
ON DEVELOPING COUNTRIES

Although it appears that developing nations can choose widely from the bountiful areas of biotechnology, each nation will, in fact, determine what will fit best within its social, cultural, and economic framework. Because many of these countries depend on their own agriculture to feed their populace, their major biotechnological thrust is likely to be in agricultural improvements.

Food crops that are better sources of nutrients, have greater yields, are more tolerant of extreme conditions, and resist disease are likely to have major effects on the food-growing regions of the world, especially in the developing countries. Simple methods to improve plant growth, such as the application of biofertilizers to crops, may require little in the way of technology and would be easy to implement. In India, for example, the introduction of earthworms and their castings (excrement) to degraded lands, along with other biofertilizers, has recovered land for agriculture.[116] Biological pest control, employing deterrent sprays produced by GMOs into which genes that code for production of natural pesticides have been cloned from plants, bacteria, or fungi, uses a technique—spraying—with which farmers are well acquainted. Disease-resistant animals, animals that can survive harsher conditions, and animals that are more efficient utilizers of feed also could have an important effect on world agriculture.

Training in tissue culture and micropropagation techniques may aid in the establishment or expansion of a locally based industry. For example, the Cycad Specialist Group of the International Union for Conservation of Nature and Natural Resources (IUCN) has suggested that local people be encouraged to raise cycads for sale from seeds or vegetative propagation in order to protect endangered plants from exploitation by commercial collectors, who pay little.[117] Such a cycad nursery program is under development in Mexico. Training programs in tissue culture and micropropagation techniques are being carried out in Costa Rica and also would be beneficial to the Colombian cut-flower industry.[118]

The products of medical biotechnology are most likely to be of immediate benefit to developing nations, especially vaccines against the major scourges of the less-developed world such as malaria, hepatitis, dengue fever, HIV, and tuberculosis[119]; the diagnostics and drugs needed to treat endemic diseases and highly infectious diseases; and the drugs and technologies that will have the widest range of applicability to increase the health of the populace. Although specialized drugs may not be major commodities in the developing nations at this time, some biotechnology firms are nevertheless optimistic. For example, Amgen's Neupogen, which is used to treat neutropenia associated with cancer chemotherapy or bone marrow transplantation (prohibitively expensive therapies) is now distributed in China. Neupogen has been approved for use in some countries for severe anemias. Epogen, human recombinant erythropoietin, is used to stimulate red blood cell production, especially in kidney dialysis patients. In

1993, Johnson and Johnson, which markets the drug in Europe and for the U.S. non-dialysis market, chalked up Epogen sales of $650 million.[120]

Many developing countries, as well as the former Eastern European bloc, have serious environmental problems that are highly amenable to bioremediation.[121] For example, much of Eastern Europe suffers from a legacy of heavy manufacturing, mining, and weapons testing without environmental controls. Poland's Vistula River is so polluted that its waters cannot be used at all. Approximately 80 percent of water samples tested from 200 river systems in the former Soviet Union showed bacterial and viral contamination levels so high that they were a threat to public health.[122] Asian rivers—even those in developed countries such as Taiwan—are among the worst in the world, containing raw sewage and industrial wastes that compromise public health and threaten entire ecosystems. Nevertheless, in many developing nations environmental cleanup has a lower priority than feeding and protecting the health of the population at large. Countries have pledged aid to the Eastern European countries, but it has not been forthcoming.[123] The World Bank just approved a $110 million loan to Russia, which will be supplemented by $90 million from the European Union, the United States, and other countries for environmental remediation.[124] Thus there is now an excellent opportunity to encourage and implement the efficient and cost-effective use of this technology.

PROBLEMS IN ADOPTING THE NEW BIOTECHNOLOGIES

Lack of Capital

For the developing countries, the new technologies will serve as a means of creating wealth[125] through products that not only can be used within the country, but also can be sold on regional or world markets. Barker,[126] however, cautions that any protective tariffs levied by these countries may induce other countries to seek substitutes for products exported by developing countries. For example, high-fructose corn sweetener, a product of the fermentation of maize, accounts for 50 percent of the U.S. sweetener market, a market that previously relied heavily on importation of cane sugar from developing countries. Furthermore, substitutes for other tropical products may become available in a trade atmosphere that discriminates against imports from developing nations. For example, according to the Rural Advancement Foundation International (RAFI), the United States is the world's largest importer of pyrethrum, a natural insecticide from the dried heads of the chrysanthemum, *Chrysanthemum cinerariaefolium*. Kenya is the world's largest pyrethrum producer; other sources are Tanzania, Ecuador, Rwanda, and Tasmania, located off the coast of Australia. If a U.S. company began to produce a genetically engineered pyrethrum product, Kenya's $75 million annual trade in the material—much of which is derived from plant micropropagation—might be destroyed.[127]

Technologies that need little capital investment but would be money-saving or produce better-quality products also need to be considered. For example, a National Research Council advisory panel suggested modernization of the production of traditionally fermented foods at the village level, using affordable technology.[128]

Developing countries do not have the capital to engage in sophisticated biotechnological research and development. Although they may have the work force—some of whom may be well trained—expensive equipment, reagents, and process control are beyond their economic means. Thus it may be preferable that organisms to be used in developing nations be researched in the more affluent countries, but manufactured (grown or maintained) in the developing countries so that they can reap the benefit of these organisms. China and India and some funded research laboratories in Africa and other parts of Asia have the trained personnel and, in some cases, the necessary equipment. In such instances, the research groups, with additional support in the form of equipment and supplies, could carry out the molecular biology research needed to produce GMOs or related products.

Except for technologies requiring only traditional skills, such as those needed to plant seeds, use of most of the new technologies will require upgrading skills of local people and extensive public education to inform the populace about the technologies. Thus the introduction of value-added, high-technology products must include educational programs.

Property Rights and Biological Prospecting

There are fears, often well founded, within the chemical and biotechnology industries that their patented materials will not be protected in developing nations.[129] Some believe that international agreements, such as the General Agreement on Tariffs and Trade (GATT), will help to assuage these fears.[130] Others see GATT as imposing systems that benefit nations in the North at the expense of the people of nations in the South.[131]

This being said, it is a fact that the world's tropical nations possess rich biotas that include especially valuable sources of medically active metabolites and natural products. Some compounds, although not yet characterized fully, are familiar in the local lore of indigenous people. But how do the more affluent nations, which tend to be in temperate climates, gain access to these riches? This question is being debated worldwide, and recent agreements, such as the one between Merck and INBio for extraction from plant material in Costa Rica, have come under criticism.[132]

Other questions include: How are indigenous people who supply knowledge, their lands (resources), and their plant matter properly compensated? How are government entities compensated, if such compensation is deemed appropriate? Some indigenous peoples who share their knowledge of native medicine with researchers and corporations that later develop these materials into drugs believe they should be rewarded for their information, in some cases with a patent. A recent review of patent law, however, concluded that this information cannot be protected by patents.[133] Likewise, it has been suggested that unique, indigenous plants be patented, but, again, naturally occurring organisms that are not products of breeding programs or any scientific genetic manipulation are not now patentable.[134] At the very least, however, these plants may be eligible for protection by the Convention on International Trade in Endangered Species of Wild Fauna and Flora (CITES) under newly proposed IUCN categories.[135] Although this convention imparts no economic rights, it gives originating countries some degree of control over who takes the plants, where the plants are sent, and what uses will be made of them. One recently published book suggested that a radical change is needed in the concept of intellectual property, putting value on culturally transmitted knowledge as well as discoveries,[136] but this is unlikely to occur in the near future. Some of these issues were addressed at the Convention on Biological Diversity (also known as the Rio Convention), but they were not spelled out clearly and none of the current agreements fully address them.[137]

In dealing with biological prospecting, also called accessing, all sides have to consider both what is fair and what is workable. Recently, a group of international Pew Charitable Trust scholars met to write ethical guidelines for bio-accessing that cover the behavior of and interactions with scientists, gene banks, and intergovernmental organizations. The guidelines propose that scientists treat indigenous peoples with respect, have local people serve as co-researchers, and ensure that the local communities receive equitable compensation for any products derived from locally collected and documented plant, microorganism, or animal-derived resources. Such guidelines will be effective only if there is a way to enforce agreements.

Although the Pew scholars may ask professional organizations to enforce member compliance, they also will append guidelines to an enforceable international treaty such as the Rio Convention. Janzen et al. have explained what a biodiversity research agreement between a researcher and "in-country biodiversity custodians" should include, but currently such arrangements vary.[138] The agreement between INBio and Merck gives the Costa Ricans cash in advance, trained personnel in the form of "parataxonomists" who can identify plants, and a percentage of sales of any products derived.[139] In contrast, the director of a herbarium in a southeast Asian country that has many unique plant species was approached by a large university from outside the region requesting that the herbarium provide it with local plant materials in exchange for vehicles and funding to pay for collecting the plants. The herbarium director, believing quite rightly that the university was taking advantage of his institution's impoverished condition, asked for a cooperative agreement between his institution, local universities, and the organization that requested the plant material, as well as some control over the material. The university was never heard from again and the herbarium director was roundly criticized by his colleagues for letting a "golden opportunity" pass. Although guidelines cannot cover all situations—one scholar involved in drafting the Pew guidelines admitted they do not cover his research situation—they may aid in reaching fair and equitable agreements. The Brazilian government is now considering an industrial property bill that could be used as a model for determining compensatory agreements between the accessors and the sources of biodiversity.[140]

An added problem in dealing with biodiversity accessing is enforcement of the Rio Convention. For example, the United States is one of the major forces for worldwide conservation, but it is not yet an official signatory of the convention. President Bill Clinton, without congressional approval, signed the treaty but with interpretive statements on Articles 16 (technology transfer) and 19 (biosafety protocols).[141] A Republican-dominated Congress is not likely to approve the initiative.

Safety and Ethical Issues

Although problems are associated with the public's perception of the safety of GMOs,[142] numerous field trials have been carried out worldwide,[143] and since 1987 field tests of more than 860 transgenic crops have been approved in the United States; at least another 250 tests have been approved in Europe since 1991.[144] Regulation and safety protocols may be accomplished with the assistance of international oversight organizations and by agreements, or through national or local laws. UNIDO's "Voluntary Code of Conduct for the Release of Organisms into the Environment" was conceived as a basic document from which a more specific code could be built.[145] Governments lacking internal expertise can call on advice from the Stockholm Institute for Environment, funded jointly by the Swedish government and the Rockefeller Foundation. Perhaps a new international nongovernmental commission on GMOs could aid countries that need assistance in formulating regulations and evaluate projects being considered for implementation within their borders.

There is concern that biotechnology-based products may lead to pressure on consumers to purchase value-added products they may not need. The Rural Advancement Foundation International worries that the addition of genetically engineered human proteins, produced by transgenic cows, into infant formula may lead the infant formula industry to undertake aggressive marketing techniques, especially in developing countries.[146] Other questions about safety and efficacy revolve around new medical technologies. Clinical trial requirements are more complex in some countries than in others, and review may be shorter in some countries, allowing a drug to enter the marketplace in Europe, for example, earlier than in the United States.[147] This in itself is not a problem, but it will become one if a drug or vaccine is unavailable in the location with the greatest need. For example, in the recent bubonic and pneumonic plague epidemics in

India, a major problem was obtaining vaccine. An effective vaccine against pneumonic and bubonic plague had been manufactured in the United States by Cutter
Laboratories, but in 1992 Cutter sold the rights to the vaccine to another company. Because FDA regulations required that the vaccine be treated as a new
product and undergo testing, it was not available when urgently needed.[148] International cooperation, and some foresight on the part of governments, should have
been able to resolve this problem long before it became an urgent one. Companies may opt for testing a product in a country with fewer controls. For example,
because the U.S. National Institutes of Health are delaying tests of a HIV vaccine
that many fear will not be effective, the manufacturers are considering carrying
out trials in Thailand.[149]

Other Obstacles

Other obstacles to the universal adoption of biotechnology projects and products are cultural, educational, economic, governmental, and infrastructural in
nature. If, for example, difficulties are encountered in delivering agricultural
products to market, no change in the qualities of those products will overcome the
infrastructural problems. In other words, there is no reason to introduce genetically engineered apples that ship better in a region where the apples rot on the
trees because they cannot be shipped to market. Introducing a complicated test kit
for clinical use by marginally trained employees will not yield the expected
public health benefits, especially if requirements such as a "cold chain" are
involved. A recent attempt to introduce clinical test kit panels into China failed
because the enzyme-linked immunoassay (ELISA) tests, although relatively
simple to use by U.S. standards, were deemed too complex and time-consuming
by the Chinese distributor.

When introducing new crops, one must be able to distribute the starter material and explain to the farmers how best to plant and grow the crops.[150] In order
to vaccinate people against disease, an infrastructure must be in place to ensure
that the vaccine reaches the people who need it. The introduction of sophisticated
technology into an area where the supply of electricity is erratic will not lead to
progress unless changes are made in the way electricity is supplied. Moreover,
complicated regulations or corrupt governments can inhibit the flow of new
technologies. For example, recently an act of the Romanian parliament was required to import a biotechnology product needed by a local area.[151] And the
INBio-Merck agreement is successful in part because the Costa Rican government
is not corrupt, but many governments foster corruption or look the other way.[152]

Finally, as noted earlier, unless the public understands both the value of and
need for advances in biotechnology, problems of acceptance of biotechnology
products will persist.[153] Thus biotechnology products and processes, even those
discovered and proposed for use in developing nations, should include not only
the introduction of the technology but also public education.

What Technologies Will Work in Developing Countries?

What will work? In 1990, a National Research Council committee produced a list of 98 plant biotechnology projects for USAID support in developing countries. The list ranged from the development of restriction fragment length polymorphism (RFLP) maps of such plants as sorghum, cowpea, and potato, to tree tissue culture and studies of the role of biotechnology in plant agriculture.[154] Fruits, vegetables, and grains with improved nutrient content and disease resistance will add significant value to agriculture. Integrated pest management, a form of the "old" biotechnology well known in some of the developing nations, could be expanded.

Marine biotechnology, including aquaculture of fish, algae, and microalgae, is a genuinely viable area for wide application in developing nations, especially in light of the severe overfishing that is occurring today. Many countries, especially those in the Pacific Rim, already have some expertise in this area, and in others expertise could be developed with the appropriate training. Marine biotechnology programs and aquaculture not only will provide food for the table, but also can develop products from natural resources.

Vaccines and pharmaceuticals that improve public health and decrease infant mortality, as well as test kits that permit screening of large proportions of at-risk populations for transmissible, even hereditary, disease will be welcomed into the markets of developing nations. In fact, nations should be encouraged to form the infrastructure necessary to develop their own vaccines, especially "orphan vaccines" for tropical diseases specific to their country. Thailand, for example, is developing its own vaccine production capability.

Programs for alternative energy sources, especially for countries that are dependent on imported fossil fuels, should be encouraged. Methane gas production, as well as the production of bioethanol and other fuels, may be an economically advantageous means of augmenting the use of fossil fuels, hydroelectricity, and nuclear power.

Environmental bioremediation can be used to introduce or upgrade public sanitation, clean polluted soil and water, and clean up toxic environments.

Finally, the development of databases, especially related to depositories of biological material, also may be important projects for international cooperation. The establishment and use of germplasm banks not only will help to preserve biodiversity but also will save food resources for future use.

CONCLUSION

Because science is international, international advisory panels, oversight groups, biodiversity consortia, research and granting organizations, and scientific societies are agents for problem solving on a global level and pooling resources across national boundaries. International organizations such as the World Bank and the United Nations, together with international treaties such as the biodiversity treaty, can sponsor the establishment of databases and networks that will allow greater international communication and cooperation. The technologies are ready for exploitation; it is the financing and the will to put these technologies into place that are needed.

ACKNOWLEDGMENT

The excellent and critical assistance of Dr. Myrna Watanabe in preparation of this manuscript is gratefully acknowledged.

NOTES

1. *Webster's Ninth New Collegiate Dictionary* (Springfield, Mass.: Merriam-Webster, 1984).

2. U.S. Congress, Office of Technology Assessment, *Commercial Biotechnology: An International Analysis,* OTA-BA-218 (Washington, D.C.: Government Printing Office, 1984).

3. U.S. Congress, Office of Technology Assessment, *Biotechnology in a Global Economy,* S/N 052-003-1258-8 (Washington, D.C.: Government Printing Office, 1991).

4. Vivian Moses and Ronald E. Cape, eds., *Biotechnology: The Science and the Business* (New York: Harwood Academic Publishers, 1991).

5. M. J. R. Nout, "Upgrading Traditional Biotechnological Processes," in *Applications of Biotechnology to Traditional Fermented Foods* (Washington, D.C.: National Academy Press, 1991), 11-19.

6. Lily E. Kay, *The Molecular Vision of Life: Caltech, The Rockefeller Foundation, and the Rise of the New Biology* (New York: Oxford University Press, 1993).

7. Barbara McClintock, "A Cytological Demonstration of the Location of an Interchange between the Non-homologous Chromosomes of *Zea mays,*" *Proceedings of the National Academy of Sciences* 16 (1930): 791-796; Harriet B. Creighton and Barbara McClintock, "A Correlation of Cytological and Genetical Crossing-over in *Zea mays,*" *Proceedings of the National Academy of Sciences* 17 (1931): 492-497.

8. M. M. Rhoades, "The Genetic Control of Mutability in Maize," *Cold Spring Harbor Symposia in Quantitative Biology,* Vol. 9 (1941): 138-144.

9. F. Griffith, "The Significance of Pneumococcal Types," *Journal of Hygiene* 27 (1928): 113-159.

10. O. T. Avery, C. M. MacLeod, and M. McCarty, "Studies on the Chemical Nature of the Substance Inducing Transformation of Pneumococcal Types. Induction of Transformation by a Deoxyribonucleic Acid Fraction Isolated from Pneumococcus Type III," *Journal of Experimental Medicine* 79 (1944): 137-158.

11. J. D. Watson and F. H. C. Crick, "Molecular Structure of Nucleic Acids," *Nature* 171 (1953): 740-741.

12. S. Cohen et al., "Construction of Biologically Functional Bacterial Plasmids *in vitro,*" *Proceedings of the National Academy of Sciences* 70 (1973): 3240.

13. U.S. Congress, OTA, *Biotechnology.*

14. Kenneth B. Lee, Jr. and G. Steven Burrill, "Biotech 95: Reform, Restructure, Renewal," Ernst and Young, Palo Alto, Calif., 1994.

15. Ibid.

16. Ibid. Unless otherwise noted, amounts are given in U.S. dollars.

17. U.S. Congress, OTA, *Biotechnology.*

18. Lee and Burrill, "Biotech 95."

19. Peter J. Russell, *Fundamentals of Genetics* (New York: HarperCollins, 1994), 313.

20. Pieter Lucas et al., "European Biotech 94: A New Industry Emerges," Ernst and Young, Brussels, Belgium, 1994.

21. Tony Going and Peter Winter, "Canadian Biotech '94: Capitalizing on Potential," Ernst and Young, Thornhill, Ont., 1994.

22. James Manuso, report given at the annual meeting of the New York Biotechnology Association, New York, N.Y., October 21, 1994; Will Mitchell, Thomas Roehl, and John Campbell, "Trends in Pharmaceutical Sales, R&D, and Profitability in the Japanese Pharmaceutical Industry before and after Ministry of Health and Welfare Pharmaceutical Reimbursement Price Adjustments, 1981-1992," unpublished manuscript, University of Michigan School of Business and Kelo University Medical School, Tokyo, revised February 14, 1994.

23. Anonymous, "World Biology Center," *Science* 266 (1994): 222.

24. M. E. Watanabe, "Hong Kong Expands Its Biotech Effort despite Eventual Chinese Takeover," *Genetic Engineering News* 10 (1990): 23.

25. Dean H. Hamer and Shain-dow Kung, *Biotechnology in China* (Washington, D.C.: National Academy Press, 1989).

26. Jon Cohen, "Bumps on the Vaccine Road," *Science* 265 (1994): 1371-1373.

27. Anonymous, "New UK Biotech Center," *Biotechnology Notes* 7 (1994): 5.

28. Rachel Nowak, "Plans for Tunisian Institute Move Ahead," *Science* 266 (1994): 359-360.

29. *BioLink: The Quarterly Newsletter of the Agricultural Biotechnology for Sustainable Productivity Project*, Vol. 1, no. 3 (1993).

30. E. DaSilva, *African Network of Microbiological Resources Centres (MIRCENs). Biofertilizer Production and Use* (Paris: UNESCO, UNDP, June 1993).

31. U.S. Department of Agriculture, "Marine Biotechnology and Aquaculture," draft report, USDA, Washington, D.C., February 7, 1994.

32. Lee and Burrill, "Biotech 95."

33. Ibid.

34. Ibid.

35. Jon Cohen, "AIDS Vaccines: Are Researchers Racing toward Success, or Crawling?" *Science* 265 (1994): 1373-1375.

36. Cohen, "Bumps on the Vaccine Road."

37. N. Regina Rabinovich et al., "Vaccine Technologies: View to the Future," *Science* 265 (1994): 1401-1404.

38. Cohen, "Bumps on the Vaccine Road."

39. Jules Musing, speech given at the annual meeting of the New York Biotechnology Association, New York, N.Y., September 21, 1994.

40. Lee and Burrill, "Biotech 95."

41. Ann Gibbons, "Childrens' Vaccine Initiative Stumbles," *Science* 265: 1376-1377.

42. Symposium on Science and Technology in Africa, Nairobi, Kenya, February 14-15, 1994.

43. National Research Council, *Plant Biotechnology Research for Developing Countries* (Washington, D.C.: National Academy Press, 1990).

44. *BioLink.*

45. Lee and Burrill, "Biotech 95."

46. U.S. Congress, OTA, *Biotechnology.*

47. Jack R. Coulson et al., "Notes on Biological Control of Pests in China, 1979," in "Biological Control of Pests in China," China Program, Scientific and Technical Exchange Division, OICD, U.S. Department of Agriculture, Washington, D.C., 1982.

48. Anonymous, "Virus Resistant Plant Remains on the Court," *Biotechnology Notes* 7 (1994): 1-2.

49. Ibid.

50. Anne Simon Moffat, "Mapping the Sequence of Disease Resistance," *Science* 265 (1994): 1804-1805.

51. John Lee Compton, speech given at the annual meeting of the New York Biotechnology Association, New York, N.Y., October 20, 1994.

52. Jocelyn Kaiser, "Lighting Up New Genes," *Science* 266 (1994): 735.

53. K. Mulongoy et al., "Biofertilizers: Agronomic and Environmental Impacts and Economics," in *Biotechnology Economic and Social Aspects. Issues for Developing Countries,* ed. E. J. DaSilva, C. Rutledge, and A. Sasson (Cambridge: Cambridge University Press, 1992), 55-69.

54. Raymond A. Zilinskas et al., "A Draft Report on the Global Challenge of Marine Biotechnology: A Status Report on Marine Biotechnology in the United States, Japan and Other Countries," National Sea Grant College Program and Maryland Sea Grant College, College Park, Md., 1994.

55. Food and Agriculture Organization, "Aquaculture Production 1985-1991," Fisheries Circular No. 815, Rev. 5, FAO, Rome, June 1993.

56. U.S. Department of Agriculture, "Marine Biotechnology and Aquaculture Report."

57. Tony Emerson, "It's Over for Fishing Here," *Newsweek,* April 25, 1994, 31-35; Anne Swardson, "Net Loss: Fishing Decimating Oceans' Unlimited Bounty," *Washington Post,* August 14, 1994, A28.

58. Zilinskas et al., "A Draft Report."

59. Kathryn Barry Stelljes, "Diagnosing the Tough Ones," *Agricultural Research* 42 (1994): 31.

60. Anonymous, "Promising Biotech Vaccine to be Tested," *Agricultural Research* 42 (1994): 31.

61. I. Y. Hamdan and J. C. Senez, "The Economic Viability of Single-Cell Protein (SCP) Production in the Twenty-First Century," in *Biotechnology: Economic and Social Aspects,* 142-164.

62. Anna Maria Gillis," Bringing Back the Land," *BioScience* 41 (1991): 68-71.

63. National Research Council, *In Situ Bioremediation: When Does It Work?* (Washington, D.C.: National Academy Press, 1993).

64. Lee and Burrill, "Biotech 95."

65. National Research Council, *In Situ Bioremediation.*

66. David T. Gibson and Gary S. Sayler, *Scientific Foundations of Bioremediation: Current Status and Future Needs* (Washington, D.C.: American Academy of Microbiology, 1992).

67. Marlise Simons, "East Europe Still Choking on Air of the Past," *New York Times,* November 3, 1994, A1, A14.

68. Going and Winter, "Canadian Biotech 94."

69. Lucas et al., "European Biotech 94."

70. Ronald M. Atlas, "Bioaugmentation to Enhance Microbial Remediation," in *Biotreatment of Industrial and Hazardous Waste,* ed. Morris A. Levin and Michael A. Gealt (New York: McGraw-Hill, 1993), 19-37.

71. Daryl F. Dwyer, "Development of Genetically Engineered Microorganisms and Testing of Their Fate and Activity in Microcosms," in *Proceedings of the Second International Symposium on the Biosafety Results of Field Tests of Genetically Modified Plants and Microorganisms,* Goslar, Germany, May 11-14, 1992, ed. R. Casper and J. Landsmann (Braunschweig, Germany: Biologische Bundesanstalt für Land- und Forstwirtschaft).

72. Steven D. Aust, "Degradation of Environmental Pollutants by *Phanerochaete chrysosporium,*" *Microbial Ecology* 20 (1990): 197-209; David P. Barr and Steven D. Aust, "Mechanisms White Rot Fungi Use to Degrade Pollutants," *Environmental Science and Technology* 28 (1994): 78A-87A.

73. P. H. Pritchard, "Effectiveness and Regulatory Issues in Oil Spill Bioremediation; Experiences with the *Exxon Valdez* Oil Spill in Alaska," in Levin and Gealt, *Biotreatment of Industrial and Hazardous Waste,* 269-307.

74. Daniel Owen, "Bioremediation of Marine Oil Spills: Scientific Validity and Operational Constraints," in *Proceedings of the Fourteenth Arctic and Marine Oilspill Program Technical Seminar, June 12-14, 1991* (Vancouver: Environment Canada, 1991), 119-130.

75. Ibid.

76. Morris A. Levin and Michael A. Gealt, "Overview of Biotreatment Practices and Promises,"

in Levin and Gealt, *Biotreatment of Industrial and Hazardous Waste*, 1-18; Ronald M. Atlas, "Bioaugmentation to Enhance Microbial Remediation," in Levin and Gealt, *Biotreatment of Industrial and Hazardous Waste*, 19-37.

77. Atlas, "Bioaugmentation."

78. D. J. Roberts et al., "Field-scale Anaerobic Bioremediation of Dioseb-contaminated Soils," in Levin and Gealt, *Biotreatment of Industrial and Hazardous Waste*, 219-244.

79. Don Comis, "Farming Ragweed and Other Plants to Clean Up Toxic Metals," Agricultural Research Service, U.S. Department of Agriculture, Washington, D.C., May 1994.

80. Anonymous, "Used Sea Sweep Being Recycled," *R&D Magazine,* October 1994, 13.

81. Anonymous, "Ag Byproducts Could Clean Wastewater," *Agricultural Research* 42 (1994): 23.

82. Jo Thomas, "Sludge Still Causes a Stink in Sunset Park," *New York Times,* October 3, 1994, B1, B8.

83. Piero M. Armenante, "Bioreactors," in Levin and Gealt, *Biotreatment of Industrial and Hazardous Waste*, 65-112.

84. Simons, "East Europe."

85. "Envirogen Awarded NSF Innovation Contract for Biodegradation of Ozone-depleting Chemicals," press release, Envirogen, February 15, 1994.

86. Geoffrey Lean and Don Hinrichsen, *WWF Atlas of the Environment,* 2d ed. (London: HarperPerennial, 1994).

87. Ibid.

88. M. D. de Jong, P. C. Scheepens, and J. C. Zadoks, "Risk Analysis for Biological Control: A Dutch Case Study in Biocontrol of *Prunus serotina* by the Fungus *Chondrostereum purpureum,*" *Plant Disease* 74 (1990): 189-194.

89. Robin Lambert Graham, Monica G. Turner, and Virginia H. Dale, "How Increasing CO_2 and Climate Change Affect Forests," *BioScience* 40 (1990): 575-587.

90. Ibid.

91. "Effects of CO_2 and Climate Change on Forest Trees," Environmental Research Lab, U.S. Environmental Protection Agency, Corvallis, Ore., April 1993.

92. Anonymous, "Halophytes May Offer Many Environmental Benefits," *R&D Magazine,* October 1994, 124.

93. Zilinskas et al., "A Draft Report."

94. Colin Ratledge, "Biotechnology: The Socio-economic Revolution: A Synoptic View of the World Status of Biotechnology," in DaSilva et al., *Biotechnology: Economic and Social Aspects*, 1-22; F. Rosillo-Calle et al., "Bioethanol Production: Economic and Social Considerations in Failures and Successes," in DaSilva et al., *Biotechnology: Economic and Social Aspects,* 23-54; M. Watanabe, "Thermophilic Biodigestion Yields a Keratinase Enzyme," *Genetic Engineering News* 12 (1992): 13.

95. "Fossil Energy Biotechnology: A Research Needs Assessment Final Report," Office of Program Analysis, Office of Energy Research, U.S. Department of Energy, Washington, D.C., November 1993.

96. M. Watanabe, "Molecular Motors Drive Multidisciplinary Research Quest," *Scientist* 7 (1993): 14.

97. National Research Council, *STAR 21: Strategic Technologies for the Army of the Twenty-first Century* (Washington, D.C.: National Academy Press, 1992).

98. Michael Freemantle, "Antigen Monolayer Electrode May Lead to Reusable Immunosensors," *Chemical and Engineering News* 72 (1994): 16-17.

99. Zilinskas et al., "A Draft Report."

100. Ibid.

101. National Research Council, *STAR 21.*

102. Ibid.

103. Ibid.

104. Watanabe, "Molecular Motors," *Scientist* 7 (1993): 14.

105. David M. Barbano, "Fact Sheet on BST," Cornell University, Ithaca, N.Y., undated; Dale E. Bauman, "Human Health Aspects of Bovine Somatotropin," Cornell University, Ithaca, N.Y., 1994.

106. Keith Schneider, "Despite Critics, Dairy Farmers Increase Use of a Growth Hormone in Cows," *New York Times,* October 10, 1994, 26.

107. Jane Rissler and Margaret Mellon, *Perils amidst the Promise: Ecological Risks of Transgenic Crops in A Global Market* (Washington, D.C.: Union of Concerned Scientists, 1993).

108. Julie Ann Miller, "Biosciences and Ecological Integrity," *BioScience* 41 (1991): 206-210.

109. Rissler and Mellon, *Perils.*

110. Casper and Landsmann, *Proceedings.*

111. UNIDO, "Voluntary Code of Conduct for the Release of Organisms into the Environment (with Annotations)," UNIDO Secretariat for the Informal UNIDO/UNEP/WHO/FAO Working Group on Biosafety, April 1992.

112. U.S. Food and Drug Administration, "Interim Guidance on the Voluntary Labeling of Milk and Milk Products from Cows that Have Not Been Treated with Recombinant Bovine Somatotropin," Docket No. 94D-0025, February 7, 1994.

113. Richard Stone, "Analysis Questions BST's Safety to Cows," *Science* 266 (1994): 355.

114. Thomas Hoban and Patricia Kendall, "Public Perceptions of the Benefits and Risks of Biotechnology," in *Agricultural Biotechnology: A Public Conversation about Risk,* ed. June Fessenden MacDonald (Ithaca, N.Y.: National Agricultural Biotechnology Council, 1993), 73-86.

115. Will Erwin, "Risk Assessment: A Farmer's Perspective," in MacDonald, *Agricultural Biotechnology,* 65-72.

116. Beena Pandey and Sachin Chaturvedi, "Vermiculture: Nature's Bioreactors for Soil Improvement and Waste Treatment," *Biotechnology and Development Monitor* 16 (1993): 8-9.

117. Dennis W. Stevenson, "Cycad Specialist Group," *Species* 21-22 (1993-1994): 102.

118. Anonymous, "Technical Training on Tissue Culture," *Biotechnology and Development Monitor* 16 (1993): 9.

119. "New Malaria Vaccine Is Effective in Mice," *New York Times,* October 11, 1994, C7; Cohen, "Bumps on the Vaccine Road"; Ruth S. Nussenzweig and Carole A. Long, "Malaria Vaccines: Multiple Targets," *Science* 265 (1994): 1381-1383.

120. Musing, speech to New York Biotechnology Association.

121. National Aeronautics and Space Administration, "Science and Society" (newsletter), Nos. 40, 41 (1994).

122. Lean and Hinrichsen, *WWF Atlas.*

123. Simons, "East Europe."

124. David E. Sanger, "World Bank Approves Loan to Help Russia Clean Up Pollution," *New York Times,* November 9, 1994, A6.

125. Ratledge, "Biotechnology."

126. R. Barker, "Scientific, Social, and Economic Implications of Biotechnology for Developing Countries," in *Biotechnology: Enhancing Research on Tropical Crops in Africa,* ed. G. Thottappilly et al. (Ibadan, Nigeria: CTA/IITA, 1992), 331-336.

127. Anonymous, "Genetic Engineering of Pyrethrins: Early Warning for East African Pyrethrum Farmers," RAFI Communiqué, June 1992.

128. National Research Council, *Applications of Biotechnology to Traditional Fermented Foods.*

129. Barker, "Scientific, Social, and Economic Implications of Biotechnology"; Andrew Pollack, "U.S. Is Shifting Trade Emphasis Away from Japan: A Focus on Rest of Asia, Latin Nations," *New York Times,* November 4, 1994, D1, D2.

130. Gabrielle Josephine Persley, "World Bank Supports Agricultural Biotech," in *Biotechnology Report* 1994/95 (London: Campden Publishing Ltd., 1994), 38-39.

131. International Development Research Centre, T*he Crucible Group, People, Plants, and Pat-*

ents: The Impact of Intellectual Property on Trade, Plant Biodiversity, and Rural Society (Ottawa: IDRC, 1994).

132. Christopher Joyce, *Earthly Goods: Medicine Hunting in the Rainforest* (Boston: Little, Brown, 1994); Kelly J. Kennedy and Charles Zerner, "What Is Equity in Biodiversity Exploration? Institutional Approaches to the Return of Benefits to Developing Nations, Communities, and Persons," paper presented to the Symposium on Biological Diversity (SWISSAID, WWF Suisse, and WWF International), Berne, October 20-21, 1994.

133. Josephine R. Axt et al., "Biotechnology, Indigenous Peoples, and Intellectual Property Rights," Congressional Research Service Report to Congress, 1993.

134. IDRC, "The Crucible Group."

135. Georgina Mace and Simon Stuart, "Draft IUCN Red List Categories, Version 2.2," *Species* 21-22 (1993-1994): 13-24.

136. Joseph Henry Vogel, *Genes for Sale: Privatization as a Conservation Policy* (New York: Oxford University Press, 1994).

137. Lyle Glowka et al., "A Guide to the Convention on Biological Diversity. Environmental Policy and Law Paper No. 30," IUCN-World Conservation Union, Gland, Switzerland, 1994; Kennedy and Zerner, "What Is Equity in Biological Diversity?"; UNCED, "Biological Diversity Convention: A Statement of Principle," *International Bioindustry Forum,* August 1993.

138. Daniel H. Janzen et al., "Research Management Policies: Permits for Collecting and Research in the Tropics," in *Biodiversity Prospecting: Using Genetic Resources for Sustainable Development,* ed. Walter V. Reid et al. (Washington, D.C.: World Resources Institute, 1993), 131-157; Kennedy and Zerner, "What is Equity in Biodiversity Exploration?"; Daniel M. Putterman, Trade and the Biodiversity Convention," *Nature* 371 (1994): 553-554.

139. Sarah A. Laird, "Contracts for Biodiversity Prospecting," in Reid et al., *Biodiversity Prospecting,* 99-130.

140. Silvio Valle, "Enabling Biodiversity," *Bio/Technology* 12 (1994): 1040.

141. Russ Hoyle, "Unfortunately, the Biodiversity Treaty Is Dead," *Bio/Technology* 12 (1994): 968-969; Vogel, "Genes for Sale."

142. "UNCED Biological Diversity Convention"; Harold A. Mooney and Giorgio Bernardi, eds. *Introduction of Genetically Modified Organisms into the Environment. Scope 44* (Chichester, England: John Wiley, 1990); Rissler and Mellon, "Perils amidst the Promise"; M. Sussman et al., eds., *The Release of Genetically-engineered Micro-organisms* (London: Academic Press, 1988); UNIDO, "Voluntary Code of Conduct."

143. Casper and Landsmann, *Proceedings.*

144. Mike Ward, "Analyzing EU and U.S. Agbiotech Field Trials," *Bio/Technology* 12 (1994): 967-968.

145. UNIDO, "Voluntary Code of Conduct."

146. Anonymous, "Biotechnology Company Will Sell Bio-engineered Human Proteins to Infant Formula Manufacturers," RAFI Communique, June 1993.

147. Roderick C. T. Scheffer and Dorine Mulder, "Regulatory Affairs," in *Biotech Opportunities in Europe: Establishing and Operating a Health Care Biotech Venture,* ed. René V. van der Kwaak and George J. M. Hersbach (The Hague: Publimarket Communications Management, 1993), 103-116.

148. Steve Sternberg, "Bottleneck Keeps Existing Vaccine Off the Market," *Science* 266 (1994): 22-23.

149. Constance Holden, "Weighing HIV Vaccine Trials," *Science* 265 (1994): 735.

150. Bertus Haverkort and Wim Hiemstra, "Differentiating the Role of Biotechnology," *Biotechnology and Development Monitor* 16 (1993): 3-5.

151. Nick Raby, speech given at the annual meeting of the New York Biotechnology Association, New York, N.Y., September 21, 1994.

152. Daniel M. Putterman, Trade and the Biodiversity Convention," *Nature* 371 (1994): 553-554.

153. Hoban and Kendall, "Public Perceptions"; Anonymous, "Canadians Delay Using BST!" *Biotechnology Notes* (U.S. Department of Agriculture) 7 (1994): 2.

154. National Research Council, *Plant Biotechnology Research for Developing Countries* (Washington, D.C.: National Academy Press, 1990).

Materials and Critical Technologies

P. CHAUDHARI

IBM Thomas J. Watson Research Center

Materials are ubiquitous; all known living creatures are made of and use materials for food and shelter, and materials are used in almost all human activity—for example, in health care, communication, transport, entertainment, and defense. Their importance has been recognized explicitly by referring to periods of human development as the Iron Age or the Bronze Age.

Over the last century, humankind not only has come to understand the properties and structure of the materials extant in nature but also has learned to combine the atomic elements to produce artificial materials. At the same time, there have been significant developments in processing the traditional materials ranging from tanning leather to producing steel. In fact, researchers have reached a level of sophistication in working with materials that is dazzling, opening new vistas in their understanding and applications. For example, a recently invented scanning microscope allows one to image individual atoms on a surface and move them one by one to desired locations. To realize this, scientists have learned to measure, control, and manipulate spatial objects that are a fraction of a billionth of a meter. In parallel with this spatial resolution, scientists can now probe, using flashes of light, phenomena that occur over times that are measured in a millionth of a billionth of a second.

These and other developments have been documented in many studies, the most extensive and far-reaching of which was published in 1989 by the National Research Council.[1] This paper, however, will not try to present a synopsis of this or any other study. Rather, it will touch on the importance of materials from the standpoints of industrial development, national choices of technologies in a competitive environment, and the institutional settings necessary for success. More specifically, it will look at some of the underlying factors that drive technologies in which materials play an integral part, using examples from the health care, computers/communication, and transportation industries; at critical technologies—how widely they are known, what this means, and how nations respond to this knowledge; and at the evolution of research and development in industries and in countries.

THE FACTORS DRIVING TECHNOLOGY

Many factors drive progress in a given technology. This section will touch on some of these factors and in the process illustrate a few of the frontiers of research and development.

Health Care Industry

Diagnostic equipment provides an extension of human senses. For example, magnetic resonance imaging (MRI), a relatively recent development, enables physicians to "see" diseased tissue. This and similar techniques for imaging the internal structure are relatively static—that is, the image is made before the surgeon operates—rather than being dynamic in the sense that the image can be made during an operation so that a surgeon can view the functioning of an organ to localize the area of disease. For example, a surgeon examines the very clear images of a brain tumor produced by an MRI and decides how much of the tumor and the tissue around it is to be removed. The surgeon's decision would be easier, however, if the brain activity surrounding the tumor could be monitored during the operation. Laboratory demonstrations have proven that this is indeed possible by mapping out the magnetic field generated by the brain as it operates. This magnetic field is very small—on the order of a billionth of the earth's magnetic field. Nevertheless, it can be measured with superconducting detectors.

This type of steady progress in medical diagnosis, driven by a constant need to improve diagnostic techniques, is largely founded on a broad base of research. The role of information technology in providing images easily comprehensible to the human eye is indispensable in medicine. The financial drive here is not so much international competitiveness as it is to supply state-of-the-art health care at manageable costs.

The technical issues at work in intelligent prosthetics, an area of health care that likely will grow into a multibillion-dollar industry over the next decade, are an understanding of the complex requirements of this technology, ranging from the materials that stimulate human nerve, muscle, bone, or cartilage, to the interface between the human body and the prosthetic. Today, this technical area is being addressed by small multidisciplinary groups in a university setting, leaving room for great technical innovation and financial growth while serving humankind. No one country has a commanding technical or financial lead.

Computer/Communications Industry

A look at the integrated circuit reveals that the level of silicon device integration in transistors has gone from 1 micron a generation ago to 0.5 microns today. The metallic lines that will control the transistor of the future will be less than 300 atoms wide. These devices of the future already are being made in the laboratory; the struggle, on an international scale, is to produce them on the factory floor at competitive prices. The financial opportunities and risks are great because no major nation can avoid facing the questions associated with semiconductor chip design and production. Their use will be pervasive—from feedback control in prosthetics to voice recognition devices used to control such mundane subsystems as locks in houses, radios, television, scooters, cars, and computers. Every family will, in some form or fashion, own a silicon chip in the near future. As for the magnitude of the industry as a whole, in India, for example, with its approximately 200 million families, easily tens of billions of dollars are involved.

Humans have evolved to communicate in certain "natural" ways, but visual imagery is perhaps the most developed in terms of the total amount of information per unit time absorbed by the mind. Displays are essential in this processing. If one assumes that one out of four people will own a display in the future, the economic impact is large—on the order of $100 billion. Thus the race to produce lightweight, energy-efficient, and visually appealing displays is under way. Perhaps the most well developed is the active thin-film, transistor-driven liquid crystal display. This industry is still evolving, with a few nations attempting to organize national programs to capture the winning technology.

These two examples illustrate some fundamental, technology-based policy issues. If every family will own a silicon chip and a display device, should a country invest in these technologies? To import might cost billions of dollars and yet to invest would require a billion dollars plus the determination to compete in these technologies on an international scale. The World Bank has a unique role in this decision making. It could, for example, provide a technical and financial assessment of the need to view information technology as an infrastructure investment for some countries and not for others. If such technology is viewed as an infrastructure investment, it is often easier for a government to invest large sums of money than if it is viewed otherwise.

Transport Industry

A car is a complex composite of diverse materials. Refinements in the choice of materials to make cars lighter (more energy-efficient) and safer will continue. But the biggest change will come from the incorporation of computers and communication systems, thereby heading toward a world in which once a destination is given to the car's computer, the car's computer/communication system not only will locate the quickest, safest, most economical route, but also will take the driver there automatically. Today's driver will have the choice, in a few decades, to be a passive passenger. But reaching this inevitable goal will require the development of technology and infrastructure.

CRITICAL TECHNOLOGIES

Do nations choose technologies based on their belief that these technologies are critical to their future? According to the *U.S. National Critical Technologies Report*,[2] the answer to this question is affirmative. The first column of Table 1, which lists the nine technologies considered critical by the United States, is followed by columns listing the technologies being targeted by Japan, the European Community, France, and Germany—some of the principal trading partners of the United States. Clearly, these countries also regard the same set or subset of these technologies as critical for their futures.

TABLE 1 Announced Foreign Targets for Critical Technologies

Technology-intensive Sector	Japan	European Community	France	Germany
Applied molecular biology	X	X	X	X
Distributed computing and networking	X			X
Electricity supply and distribution	X	X		X
Flexible integrated manufacturing	X	X	X	
Materials synthesis and processing	X	X	X	X
Microelectronics and optoelectronics	X	X	X	
Pollution minimization and remediation	X		X	
Software	X	X		X
Transportation	X	X	X	X

NOTE: Some member nations of the European Community (EC) have announced different targets than the EC collectively.

SOURCE: U.S. Office of Science and Technology Policy, *U.S. National Critical Technologies Report* (Washington, D.C.: OSTP, 1993).

This then leads to a second question: How widespread is the knowledge of these critical technologies? There are many ways to demonstrate that the set of critical technologies—whether they are identified by the names in Table 1 or whether they are one level higher or lower in technology hierarchy—are widely known. For example, Table 2, which shows by country the R&D investment in the United States, and Table 3, which reveals the patent activity by country in the nine critical technologies, suggest that the trading partners not only actively search for knowledge in the United States, but also participate in generating it. Intellectual property protection is highly emphasized in a competitive environment in which knowledge is pervasive, providing for both the defensive and active protection of technology. It also can be a key earner of money for technologies of the kind described above. In short, nations choose critical technologies to emphasize and organize around from a set of technologies that are widely known.

INSTITUTIONAL EVOLUTION

Given that knowledge about critical technologies is widespread and that within a given critical technology its needs are widely identified, what do different nation-states need to do to optimize the flow of knowledge from R&D to the manufacturing floor? When knowledge is widely and easily available, as it is increasingly because of growing worldwide R&D spending and communications, the extent to which an organization needs to undertake R&D is affected. The ratio of R&D to the search for and acquisition of technology (S&A)—an essential activity in a competitive environment—has a characteristic dependence on time for a given industry and can be used to group the progress of different countries in different industries. Thus some countries or organizations need to increase their R&D, while others need to increase their S&A. This is perhaps best illustrated by a schematic representation of the ratio of R&D to S&A as a function of time for the United States, Japan, and India and China (Figure 1). After World War II, the United States dominated industrial R&D in such areas as automobiles, electrical machinery, and materials (steel). But as knowledge about the technology base of these industries spread to the rest of the world, the ratio of R&D to S&A (R/S) declined in the United States, and U.S.-based industry increasingly sought technical know-how from outside its own R&D organizations. In Figure 1, the automobile, electrical machinery, and materials industries are at the lower plateau of R/S; the pharmaceutical and environmental industries are at the top plateau; and the computer/communication industry is rapidly moving from the higher to the lower plateau. This curve is characteristic of all technology-based industries. If it is accepted that such a trend exists, then the management of R&D depends on the nature and evolutionary state of a particular industry and country.[3]

In contrast to the United States, Japan had an almost nonexistent industrial R&D base after World War II (Figure 1). Japanese industry therefore searched

TABLE 2 Research and Development Investment in the United States by Selected Countries

Technology-intensive Sector	United Kingdom	France	Germany	Japan	Korea	Taiwan	All Others	Totals (billion U.S. dollars)
Applied molecular biology	●	●	○					75
Distributed computing and networking			○	○	●			29
Electricity supply and distribution								3
Flexible integrated manufacturing			●	○				3
Materials synthesis and processing			●	○				12
Microelectronics and optoelectronics			○	○	●		●	67
Pollution minimization and remediation				●				3
Software				●				28
Transportation				●				5
Total investment (billion U.S. dollars)	13	12	29	128	10	0	30	222

NOTE: ● = receives heavy emphasis relative to emphasis by other countries; ○ = receives some emphasis relative to emphasis by other countries; blank = receives little emphasis.

SOURCE: U.S. Office of Science and Technology Policy, *U.S. National Critical Technologies Report* (Washington, D.C.: OSTP, 1993).

TABLE 3 Patent Activity by Country in Technologies Defined as Critical

Technology-intensive Sector	United States	Japan	Germany	United Kingdom	France	Taiwan	Korea	Total (million U.S. dollars)
Applied molecular biology	●		●	●	●			8,703
Distributed computing and networking	○	●		●	○		●	8,044
Electricity supply and distribution	○	○	○	○	●	○	●	11,799
Flexible integrated manufacturing	○		●	○	○	○		7,168
Materials synthesis and processing	●	○	●	○	○	○		9,986
Microelectronics and optoelectronics	○	●	●	○	○	●	●	7,206
Pollution minimization and remediation	○		●	○	○	○		4,437
Software	○	●		○		○	●	2,230
Transportation		○	●	○	●	○	●	5,022

NOTE: ● = higher than average level of patenting activity in this sector; ○ = average level of patenting activity in this sector; blank = below average patenting activity in this sector.

SOURCE: U.S. Office of Science and Technology Policy, *U.S. National Critical Technologies Report* (Washington, D.C.: OSTP, 1993).

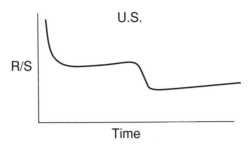

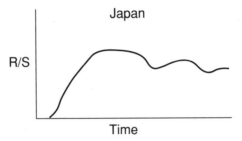

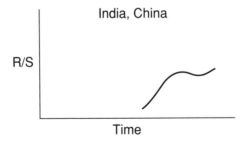

FIGURE 1 Schematic representation of the ratio of research and development (R) to search and acquisition (S) as a function of time for the United States, Japan, and India and China.

for that knowledge. Their ratio of R/S, largely stemming from a large S, was low in this time frame. As their industries reached competitive levels, the R/S ratio increased by raising R until a steady R/S ratio was reached.

Countries such as India and China are evolving differently. Although there is a large cultural bias toward research, there is almost no significant industrial R&D that attempts to seek this knowledge and turn it into competitive products. The search and acquisition in these countries are not for knowledge but rather for systems that can be transplanted onshore. This process, while profitable onshore, rarely leads to competitive products on a worldwide basis. The R/S curve for China or India is therefore shown as ramping up later in time compared to, for example, Japan. The extent of this "lateness" is determined by the ability and determination of Chinese or Indian industry to be competitive in selected technologies and, of course, by government policies on the need for such competitiveness.

In summary, while materials are necessary for development, they must be viewed as part of a national technology strategy, and care must be given to ensuring that the world environment and the ability of local institutions to sustain, develop, and utilize the know-how are understood.

NOTES

1. National Research Council, *Materials Science and Engineering for the 1990s: Maintaining Competitiveness in the Age of Materials* (Washington, D.C.: National Academy Press, 1989).

2. U.S. Office of Science and Technology Policy, *U.S. National Critical Technologies Report* (Washington, D.C.: OSTP, 1993).

3. P. Chaudhari, "Corporate R&D in the United States," *Physics Today* (December 1993): 39-40.

Information Technology for Development

JOHN S. MAYO
President Emeritus, AT&T Bell Laboratories

Recent developments in information technology will enable all countries—and especially the developing nations—to leap into the Information Age. Happily, no longer must these countries watch the advances in communications and information networking that they are seeking move forward one step at a time, as might have been the case in the past. With this in mind, this paper will delve into the possible impacts of the available information technology on developing countries. But any discussion of the role that information technology might play in such countries first must be placed in context by examining the forces propelling the emerging multimedia communications revolution and the evolution of the information superhighway, including AT&T's vision of it.

THE FORCES DRIVING THE MULTIMEDIA REVOLUTION AND THE EVOLUTION OF THE INFORMATION SUPERHIGHWAY

The key underlying information technologies are the prime drivers and the major enablers behind the emerging multimedia communications revolution and the paving of the information superhighway—as well as a host of other advances that together are changing the way in which people live, work, play, travel, and communicate. Because these key information technologies are changing the work and home environments, they also are helping to address consumer needs. In fact, the more they can do, the more new products and services the consumer wants,

---

Dr. Mayo addressed the symposium via video conferencing from AT&T in Basking Ridge, New Jersey.

producing an upward spiral that has lasted over three decades and will surely last at least one or two decades more.

But what are these key underlying information technologies? They are silicon chips, computing, photonics or lightwaves, and software. The technology capabilities have been doubling every year in a number of such domains—for example, in computing and photonics—and doubling every 18 months in silicon chips. Even software—once a "bottleneck" technology because of quality and programmer-productivity problems—is beginning to advance rapidly in such major areas as telecommunications because of advanced programming languages and reuse of previously developed software modules. Such modules contribute to programmer productivity because they can be used in more than one project, and they improve quality because they have been tested.

Perhaps the most widely known example of technology advancements is the explosive growth in the power of silicon chips—one measure of which is the number of transistors that can be crammed onto a chip the size of a fingernail. This number, now in the millions, is moving steadily toward known physical limits. In the early part of the next century, today's familiar solid-state devices may mature, with transistors measuring about 400 atoms by 400 atoms each—the smallest such transistors likely to operate reliably at room temperature. The new frontier, then, will not be in making the devices smaller, but in using creatively and economically the vast increase in complexity and power made possible by this remarkable technology.

The amazing progress of silicon chips forms a microcosm of the broad thrust of information technology and all the associated forces that are leading to the multimedia communications revolution and the evolution of the information superhighway. But what progress is being made in the related driving forces, and what impacts are they having?

After the invention of the integrated circuit, every time the number of transistors on a silicon chip increased by a factor of a thousand, something had to be reengineered—that is, something had to be radically changed or improved because it was a new ball game. As researchers headed toward the first thousand-fold increase, the reengineering took the form of changing all of AT&T's design processes, which had been based on discrete components. When the milestone of a thousand transistors per chip was attained, the new digital circuitry was used by AT&T to reengineer its products from analog to digital, as did many other industries. This early progress toward digital products, made possible by silicon chips and software, brought about the digitalization of most systems and services—both domestically and, more and more, globally—creating a powerful force that is driving the information industry toward multimedia communications and the information superhighway.

When, about a decade ago, researchers approached the milestone of a million transistors per chip, powerful microcomputers became possible, along with all the periphery related to them and the needed software systems. All this resulted in an explosion of advanced communications services, forcing the antitrust process that led to the reengineering of AT&T: from a company that provided largely voice and data-on-voice telecommunications services to a company focused on universal information services. The theme of universal information services is voice, data, and images anywhere, anytime, with convenience and economy. Such advanced services, provided on an increasingly intelligent global network, constitute the beginning of multimedia communications, now emerging as the revolution of the 1990s and beyond.

In this era of yet another thousand-fold increase in transistors per chip, reengineering has extended beyond AT&T and toward the merging of the communications, computer, consumer electronics, and entertainment industries. The bringing together of these four industries has started out in the obvious ways— that is, through joint projects, joint ventures, mergers, acquisitions and some new start-up companies. This reengineering of the information industry appears to be the next to the last step in the information revolution brought on by the invention of the transistor.

The last step, and one that may go on forever, is the reengineering of society—of how people live, work, play, travel, and communicate—creating a whole new way of life. For example, it will change education through distance learning and school at home; it will change work life through virtual offices and work at home; and it will diminish the need for people to transport themselves elsewhere for work or such routine tasks as visiting and shopping. But social change as well as technology will be needed to make many of these changes happen.

Another driving force toward multimedia communications and the information superhighway is the worldwide push toward common standards and open, user-friendly interfaces that will encourage global networking and maximum interoperability and connectivity. Photonic or lightwave transmission facilities, for example, will be based on the evolving international standard known as SDH or synchronous digital hierarchy. Because SDH defines standard network interfaces, service providers and customers will be able to use equipment from many different vendors without worrying about compatibility. This will facilitate the upgrading of existing networks and the construction of new networks on a worldwide basis. SDH also will provide efficient transport of broadband services and will simplify networks. Similar standards in domestic networks will enable digital communications to the workplace and home and will make possible high data-rate services.

The broadband integrated services digital network or B-ISDN is a new digital format as well as an international standard that supports such multiple services as voice, data, and new video services over lightwave transmission facilities. This development could introduce an exciting new era in global communications networking as equipment vendors and service providers adopt compatible standards to provide sophisticated high-bandwidth, or high-information-capacity, services. B-ISDN is currently defined at interface rates of 155 million bits per second and 622 million bits per second.

At present, the force pacing behind the multimedia and information super-highway revolution is not so much the technology as it is marketplace demands. For the greater part of this century, the user willingly accepted whatever techno-logical capabilities were available. Thus the telecommunications industry was supplier-driven, and suppliers managed the evolution of the industry and the information highway. But eventually the technology became so rich that it made many more capabilities available than the user could accept—that is, developers were able to design a lot more products and services than customers were willing to pay for. That marked the transition from a supplier-driven industry to today's customer-driven industry—from supplier push to marketplace pull. Globally, the transfer and assimilation of information technology are combining with political and regulatory forces—such as the move toward the privatization of telecommu-nications in both developed and developing countries—to result in the growth of ever-stronger competition in the provision of communications products and ser-vices. Such emerging competition is another force driving the evolution of both multimedia communications and the information superhighway. Moreover, pub-lic policy is being challenged—not just in the United States but also globally—to provide a framework within which that evolution can occur with full and fair competition for all players.

THE MULTIMEDIA REVOLUTION

The pursuit of multimedia is placing social pressures on the evolution of the information superhighway both in the United States and around the world. But what exactly are *multimedia*? The term refers to information that combines more than one medium, where the media can include speech, music, text, data, graph-ics, fax, image, video, and animation. AT&T tends to focus on multimedia prod-ucts and services that are connected over a communications and information network. Examples of such networked multimedia communications range from videotelephony and video conferencing; to real-time video on demand, interac-tive video, and multimedia messaging; to remote collaborative work, interactive information services such as electronic shopping, and multimedia education and training. Eventually, advanced virtual reality services will enable people to indi-rectly and remotely experience a place or an event in all dimensions.

Public-switched networks—or information highways—can presently accom-modate a wide array of networked multimedia communications, and the evolu-tionary directions of those networks will enable them to handle an increasingly vast range of such communications. Moreover, there is a potentially vast market for multimedia hardware and supporting software. Although actual projections differ widely, the most commonly quoted projection for the total worldwide market for multimedia products and services is roughly $100 billion by the year 2000.

AT&T is playing a major role in facilitating the emerging multimedia revolution—as a service provider, as a provider of network products to local service providers, and as a provider of products to end users. These are familiar roles for AT&T, but it also is studying another, perhaps less familiar, major role in relation to the multimedia revolution: that of "host" for a wide variety of digital content and multimedia applications developed by others. Hosting is a function that connects end users to the content they are seeking—that is, it provides easy, timely, and convenient access to personal communications, transactions, information services, and entertainment via wired and wireless connections to telephones, hand-held devices, computers, and eventually television sets. Independent sources of this digital content eventually will range from publishers to large movie studios to small cottage industry software houses.

The role of host illustrates one of the key challenges of the information superhighway because openness of critical interfaces and global standards are vital to the complex hosting function. The entertainment industry, for example, must have software systems that are compatible with those of the hosting industry, and these software systems, in turn, must be compatible with those of the communications and information-networking industry, which then must be compatible with those for the customer-premises equipment industry.

The tremendous growth in available information and databases will then stimulate the need for personal intelligent agents. Software programs activated by electronic messages in the network, these "smart agents" find, access, process, and deliver desired information to the customer. In fact, they can perform many of the time-consuming tasks that have discouraged a number of users from taking advantage of on-line services and the emerging electronic marketplace. One feature of AT&T's recently announced enhanced network service, AT&T PersonaLink Services, these smart agents can make shopping for the best mortgage, or finding the best new car deal, or finding out which store has a particular sought-after item much easier by avoiding the people at the interface who add negative value. For example, a replacement part is needed, but two calls to the store that might have it produce no satisfactory response. A trip to the store and a wait in line then produce a salesperson who queries the store's database and says, "We don't have it in stock." A smart agent could have queried the store's database and saved the store and the customer a big investment in a zero-revenue operation. There was never a problem with the database; the problem was the people who were inadvertently in the way of the customer's ability to access it— adding negative value but diligently trying to do their jobs. A smart agent simply could have done it better.

On an even more personal level, people who are geographically apart in the age of multimedia communications will not, for example, just play games together over networks; they will visit and build their relationships. Consumers and business associates will seek new relationships based on "telepresence," a new type of community, and a social experience independent of geography. This potential for interactive networks is quite unlike that found in the proposed availability in the United States of 500 preprogrammed TV channels; rather people will have the freedom to choose any subject or service from the intelligent terminals in their homes and offices. Indeed, they will be able to network clusters of friends or associates to enjoy such services as a group.

Networked multimedia communications will dramatically change the nature of work and therefore will have a broad impact on business—first in developed nations and eventually in developing nations. Video conferencing, for example, will enhance productivity, save time, and reduce travel. And current developments in multimedia telephony are making the possibility of remote collaborative work more and more realistic. In a few years, for example, a person working in real time with colleagues or suppliers in branch offices in New York, Hong Kong, Paris, and Sydney could accomplish the combined task of producing printed materials, presentation slides, and a videotape introducing a new product line.

AT&T'S VISION OF THE INFORMATION SUPERHIGHWAY

As noted earlier, the pursuit of multimedia communications is driving social issues related to the evolution of the information superhighway. AT&T envisions that the information superhighway will bring people together, giving them easy access to each other and to the information and services they want and need—anytime, anywhere. According to this view, the information superhighway is a seamless web of communications and information networks—together with other elements of the national information infrastructure, such as computers, databases, and consumer electronics—that will put vast amounts of information at the fingertips of a variety of users. The information superhighway is, quite simply, a vast interoperable network of networks embracing local, long-distance, and global networks; wireless; broadcast and cable; and satellites. In addition, the information superhighway embraces the Internet, as well as the test beds[1] associated with the High-Performance Computing Initiative[2]—such as the experimental communications network known as the Blanca test bed with which AT&T is associated. The information superhighway is not a uniform end-to-end network developed and operated by government or any one company. It is the totality of networks in the nation, interconnected domestically and globally. And it is an important part of the evolving global information superhighway.

THE IMPACTS OF TECHNOLOGY TRENDS
ON DEVELOPING NATIONS

Advanced information technology trends, multimedia communications, and the information superhighway will have a variety of broad, beneficial social impacts on developing nations. Advanced communications, growing in ubiquity, could slow the migration of rural people to urban areas—a traditional problem in such countries as the People's Republic of China. For example, people living in rural areas would be less inclined to move to the cities if advanced communications systems gave them access to jobs and sophisticated social services where they already live. In the United States, the pervasive communications infrastructure has enabled information-intensive businesses to flourish anywhere in the country. The information superhighway even could alleviate congestion and commuter traffic pollution in cities by making telecommuting possible—by bringing good jobs to people wherever they are. The work-at-home movement is gaining momentum, and trials with certain kinds of jobs show that employees can be even more highly productive without leaving their homes. One side benefit here is reduced costs for urban office space.

Information technology also could revolutionize education and eliminate differences in quality between rural and urban education systems by enabling a limited number of the very best teachers and professors to reach huge numbers of students. Both students and teachers could be located practically anywhere in "virtual" classrooms, and they could enhance learning by accessing multimedia network databases on a great variety of content areas.

Information superhighways also could revolutionize medical care by helping to deliver high-quality medical care far from large population centers. Advanced communications would permit frequent meetings between rural health workers and physicians located in more populated areas. The same capability would permit direct doctor-to-patient consultation and follow-up.

Advances in information technology also are stitching together a truly global society and a global economy in which developing nations would be able to participate fully. Peoples and countries would be able to retain their ethnic and cultural identities, yet at the same time communicate, transact, and interact seamlessly across geographic and political boundaries. Within political boundaries, a modern information infrastructure would help to strengthen the ties that hold a nation's people together. In a large country such as China, for example, the huge distances between cities and regions and the enormous complexity of regional dialects have made communication among the Chinese people exceptionally difficult. Thus the information superhighway could help to lessen both the obstacle of distance and the barrier of language. Information technology also will eventually make possible real-time translation of languages (speech in one language is automatically translated into another language and vice versa).

In addition to these social impacts, the key information technology trends, multimedia communications, and the information superhighway will have some broad public policy impacts on developing nations. For example, in general, investment in communications infrastructure contributes significantly to a nation's overall economic development. Fortunately, the new technologies in which developing countries would be investing are becoming more and more cost-effective, and in choosing the technology path that will move them most directly into the Information Age, these countries will have the opportunity to "leapfrog" many of the older technologies that preceded today's advanced network systems—for example, to install glass fiber in local distribution networks. Indeed, the technology is available to actually "jump-start" a developing nation. For example, cellular radio can provide telephony almost overnight and serve large markets while the fiber-optic infrastructure is put in place. But any such investments in physical infrastructure also will require a heavy investment in the development of the human infrastructure. The global leaders of the twenty-first century will be those countries that have invested not only in the right technologies, but also in the intellectual growth of the people who will use them.

Information technology is vital as well to economic reform and development and to attracting and meeting the needs of foreign investors. In the area of financial management, for example, information technology could enable a country to move away from a cash economy to one in which electronic transactions not only are faster, but also provide much greater visibility into economic activity.

Finally, information technology would both facilitate and complicate the job of governing. It would facilitate by making available to decision makers vastly expanded resources of timely information. And it would complicate by greatly expanding the numbers of people who would be informed about important issues and who inevitably would want to play a role in deciding them.

In the United States, the government has played a crucial role in nurturing rapid technological progress, as well as the rapid application of new technologies in the marketplace. In the communications sector, for example, the government has established a clear set of national objectives such as universal service, technological leadership, and broadband capability into all population centers. The government also has created a strong, independent regulatory structure designed to ensure that private companies serve the public interest in a fair and competitive marketplace, although there is still a way to go toward genuine and effective competition in the local exchange. Many, if not most, developing nations are still evolving their policies, laws, and regulations governing the communications industry—a very important task.

In summary, rich information technology, the worldwide push toward global standards, ever-increasing customer demands, and growing global competition are the key forces driving the emerging multimedia communications revolution and the evolution of a information superhighway—developments that promise a broad range of Information Age benefits to virtually every citizen of the United States. They also promise to extend these Information Age benefits to virtually every citizen of the world, including the developing nations.

NOTES

1. A test bed is a technical trial of leading-edge technology in order to evaluate the technology and its application.

2. The High-Performance Computing Initiative is a U.S. government initiative that funds a number of programs aimed primarily at improving computing.

Broadened Agricultural Development: Pathways toward the Greening of Revolution

RICHARD R. HARWOOD
Michigan State University

The challenge for the coming decades of sustainable agricultural develop-ment is to transform existing farms (agricultural businesses) into new industries that meet evolving human needs, use resources efficiently, and are environmen-tally stable. Over the next 25 years, efforts to provide sustainable agricultural development will be confronted with a massive increase in the world's popula-tion (perhaps a doubling), which, when coupled with economic growth, will greatly increase the demand for goods and services generated from the land and water base. Although the world's food supply will be plentiful and will see only modest price increases over the next decade, after that as yet undiscovered tech-nologies and new resources must be brought on line to sustain growth.

Also over the next 25 years, global trade patterns, together with ongoing development, will open markets for industries located wherever production re-sources are available. For agriculture, the emergence of biological technologies for the transfer of the ability to produce new industrial precursors, essential oils, and flavor compounds into agricultural crops will open many new markets and create new agricultural industries. These new industries will permit production of more products directly from solar energy, helping the transition to a solar economy, but at the same time they will put more pressure on land resources because of the increased demand for additional products from agriculture.

The "frontier" period in freshwater and marine fisheries has ended; natural supplies of fish and their products are declining while global demand is increas-ing. Thus as limits are placed on harvest from native stocks in an effort to optimize and sustain natural productivity, new industries devoted to "fish farm-ing" will be required to meet growing demand. Also during the next decade the

145

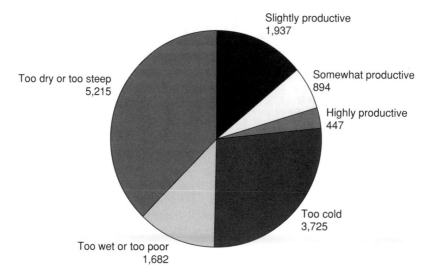

FIGURE 1   Agricultural productivity of land worldwide (million hectares). SOURCE: Rockefeller Foundation, *1993 Annual Report* (New York: Rockefeller Foundation, 1994).

earth will have reached the end of its "frontier" period in the harvest of virgin forest. The forest industry, then, will have to undergo transformation on a global scale to meet the growing demand for its products. As for water, demands will increase, with ever-greater emphasis on quality. Its value (and cost) will therefore continue to increase.

Overall, the environmental interactions among geographical areas and between industrial and human activities within areas will continue to intensify. Industry will be under increasing pressure to control outflow to the environment during production and to recycle the post-consumption product or materials. The problem of control of materials during production is particularly relevant to agriculture, which, to be economical, must manage pests and mobilize nutrient movement at high rates.

As for the land itself, the next 25 years will see a leveling off in the availability to agriculture of the land most productive for cultivated crops, followed by a decrease. The most productive agricultural areas, on a global basis, are only a small portion of the total land base, accounting for 447 million hectares out of a total 13,900 million hectares, or 3.2 percent (Figure 1). The "slightly" and "somewhat" productive lands add another 1,087 million hectares.[1] With advances in technology, some of the soils with "lower potential" can be made productive and converted to arable farming, especially in meeting the growing demand for products from perennial crops which do not require frequent soil tillage.

From 1950 to 1981, the global grainland area increased about 0.7 percent a year, while from 1981 to 1992 the grainland area decreased at an annual rate of

0.5 percent. This decrease is caused by many factors, including lower world prices for grains, the conversion of land to nonagricultural uses, and the transfer of the less productive lands from cereals to more suitable crops. Irrigated land in developing countries increased at a rate of 2.17 percent a year from 1961 to 1971, 2.09 percent a year from 1971 to 1981, and 1.24 percent a year from 1981 to 1990.[2] The most optimistic outlook for an eventual increase in irrigated area in developing countries is 59 percent above present acreage,[3] but also noteworthy is the fact that the per hectare costs of such irrigation development increase as the quality of land and the availability of water decrease.

The pockets of hunger that continue in developing countries caused by the plethora of social, political, climatic, and other disturbances will require globally available stocks of food. The poverty evident in both rural and urban areas is, of course, an ongoing problem. Rural poverty interacts with agriculture on several fronts. Its greatest long-term impact is seen in the overexploitation and degradation of fragile soils by production enterprises geared toward short-term human survival rather than optimal long-term productivity. The land areas in developing countries that are receiving the greatest portion of increases in rural population are those areas having lower soil and water availability. With their increased fragility, these poorer soils have far less tolerance for stress before reaching an irreversibly degraded phase (Figure 2), and many if not most of the less-produc-

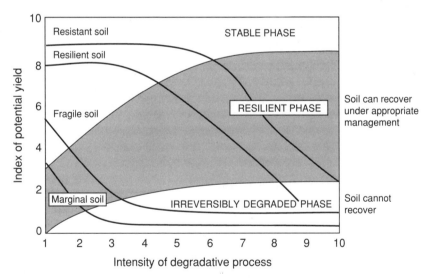

FIGURE 2 Responses of resistant, resilient, fragile, and marginal soils to stress (see note 4). SOURCE: Adapted, with permission, from R. Lal, G. F. Hall, and F. P. Miller, "Soil and Degradation: I. Basic Processes," *Journal of Land Degradation and Rehabilitation* 1 (1989): 5-69. © 1989 by R. Lal, G. F. Hall, and F. P. Miller. Reprinted by permission of John Wiley & Sons, Ltd.

TABLE 1 Prevalence of Poverty in the Developing World, 1990 and 2000

Region	Population below Poverty Line (%)	
	1990	2000
South Asia	49.0	36.9
East Asia	11.3	4.2
Sub-Saharan Africa	47.8	49.7
Middle East and North Africa	33.1	30.6
Latin America and the Caribbean	25.5	24.9
All developing countries	29.7	24.1

NOTE: The poverty line is defined as an annual income per capita of $370 in 1985 purchasing power parity dollars.

SOURCE: Reprinted, with permission, from World Bank, *World Development Report 1992* (New York: Oxford University Press, 1992). © International Bank for Reconstruction and Development.

tive soils fall into the "fragile" and "marginal" categories as defined by Lal.[4] Most of these lower-potential soils will not permanently support cultivated crops on large, contiguous areas. Annual crops such as cereal grains are not highly responsive to inputs when soil quality and water are the primary limiting factors.

Millions of poor people depend on these low-productivity areas for their subsistence. While the overall percentage of poor people in developing countries is projected to decrease during the decade of the 1990s (Table 1), the combination of population growth and migration to low-resource areas may actually cause major population growth in those areas. In the mid-1980s nearly half of the developing world's poorest people—some 370 million—lived in low-potential rural areas (Table 2). But development in high-resource areas and subsequent outmigration from poor areas will not, by themselves, solve the poverty problem in the near term for most of these people. Sustainable development must include optimizing resource use in these low-productivity areas through use of appropriate production strategies.

THE BROADENING (AND GREENING) OF THE AGRICULTURAL DEVELOPMENT PARADIGM

Against this rapidly changing global development picture the public demands on agriculture (the development paradigm) are evolving. Ironically, the quality of life of an increasing proportion of the earth's inhabitants depends on the structure and activities of nearby agriculture at the same time that the number of people making their living from farming is decreasing. Public agendas for

TABLE 2 Distribution of the Poorest of the Poor within Low- and High-Potential Areas, Mid-1980s (millions)

Region	Rural Areas		Urban Areas
	Low Potential	High Potential	
Asia	265	198	83
Sub-Saharan Africa	71	69	16
Latin America	35	12	31
All developing countries	370	277	131

NOTE: The poorest of the poor are defined as the poorest 20 percent of the total population of all developing countries.

SOURCE: Reprinted by permission of Transaction Publishers from H. J. Leonard, "Overview—Environment and the Poor: Development Strategies for a Common Agenda," in *Environment and the Poor: Development Strategies for a Common Agenda*, ed. H. J. Leonard and contributors (New Brunswick and Oxford: Transaction Books, 1989). © 1989 by H. J. Leonard; all rights reserved.

agricultural change will continue to broaden as social and environmental dimensions are added to economic demands. The breadth of the public agenda depends on three factors: (1) food sufficiency—if food is scarce or the supply is insecure, the agenda will focus narrowly on production, as in the 1960s during the early days of the green revolution when new rice and wheat varieties were combined with the rapid development of rural roads, electrification, irrigation systems, and production credit using models defined by such development giants as Art Mosher;[5] (2) the degree of a country's political and social pluralism—centrally-controlled and planned economies and societies tend to have narrow agendas; and (3) the overall affluence of the majority of people. Superimposed on all of this is the "global" agenda of the various development agencies and institutions, both large and small, emanating largely from the developed countries. As a result, an environmental or social ethic may be forced on a developing nation well ahead of its evolution at the public level. Finally, in an increasingly populous and interdependent world, the agricultural agenda will inevitably broaden to include ever greater levels of: human utility—measured in terms of production, employment, safe food, dependability, recreation, green space, and a host of other factors; efficient use of resources—measured as land use, return on investment, and use of inputs; preservation of nonrenewable resources, including soil, water, and genetic diversity; environmental impacts favorable to humans and most other species; and macro structure in harmony with local and national economic, social, and political goals. This broadened agenda is spelled out in great detail in the *Agenda 21* report of the 1992 UN Conference on Environment and Development (UNCED).[6]

One might debate the rate of change called for in this agenda, but its overall direction cannot be altered significantly. Most development agencies, including the international agricultural research centers, have reviewed their programs and developed detailed responses.[7] Individual centers such as the International Board on Soil Research and Management (IBSRAM) have responded with specific reference to their subject areas,[8] and these responses are quite typical of the dynamic changes under way.

TECHNOLOGICAL BREAKTHROUGHS

Because the world is facing a shrinking land base and growing demand for agricultural products, the output per unit area of food and feedgrains, as well as starchy vegetables, must more than double over the next 25 years. While there is considerable scope for increasing yields within the existing genetic potential, scientific breakthroughs will be needed to fully achieve the required yields. The recently announced achievement of an up to 30 percent increase in the genetic yield potential of rice is exactly the kind of progress needed. But many more years of highly complex (and expensive) research and development will be necessary to bring this new rice technology to farmers' fields. And, just as important, yet to be developed are the production systems for managing pests, water, and the high nutrient flows needed to achieve such yields without additional pesticide, fertilizer, and plant or animal waste loss to the environment.

Yields of rice in Asia, when adjusted for climate, average 57 percent of the present genetic yield potential of 8.0 tons per hectare. But the combination of pest management, disease control, and soil quality and nutrient management technologies that are required to close the yield gap are not yet available. Most important, the sustainable yield potential must be increased. Yet it is not feasible to expect average production to rise to more than perhaps 70-80 percent of maximum potential yields because farmers are unable to control the many yield-reducing variables to the extent that researchers can.

The rapidly evolving science of production ecology is giving new insight into mechanisms for better management of biological process, which should dramatically help in this regard. Researchers' abilities to manage soil microbial populations to better synchronize seasonal nutrient flux with crop demand, to manage genetic shifts in pest resistance to control measures, and to genetically engineer new characteristics into key crop-influencing organisms are but a few examples.

The genetic transfer of a plant's capacity to produce essential oils, industrial products, medicines, or other useful products represents a new wave of scientific breakthrough. The ability of plants to produce a chemical precursor for biodegradable plastic is an excellent example. Because this chemical is linked to the production of carbohydrate, this trait will be transferred first to such crops as sugar beet. If an investment were made in transforming oil palm into a plastics producer, a new industry could form in the humid tropics, where much of the nonarable land is ideally suited to sustainable tree crop production. If land tenure, credit schemes, and a development strategy for an appropriate scale of farm enterprise were in place, such an industry could have a major impact on both the productivity of the land resource and rural poverty.

In some cases, the new technologies, such as the genetic transformation of fungi or bacteria to produce high-value products, may move traditional agricultural products from a land-based to a "factory"-based system. Among other products, chocolate and vanilla production may be affected, with potentially disastrous social and economic consequences for the people and countries now producing and exporting these items. Ghana, for example, now earns a major portion of its foreign exchange from exports of cocoa and will be forced to shift a significant part of its agriculture toward alternative products.

RAPIDLY EVOLVING AGRICULTURAL INDUSTRIES

Several industries such as fisheries, farm forestry, and selected animal enterprises are evolving rapidly on a global scale to meet changing market demands. In Indonesia, because the harvest of virgin forest has passed its peak, the price of logs of certain timber species has risen by 20 percent a year.[9] This is prompting a significant shift in Indonesia's timber industry toward land management and timely production rather than continued reliance on native stands of trees. Private forest land held on long-term lease is beginning to encroach on what was once the exclusive plantation domain of the government.

The "industrialization" of swine and poultry production in Asia over the past decade illustrates the direction that fish and other animal production is taking. Nonruminant animals and fish, in particular, respond dramatically to a combination of quality genetic stock, control of disease and parasites, and proper nutrition. In fact, productivity per animal can be increased up to fivefold over that of animals foraging untended in a village-level environment. But while vertical integration of feed, veterinary care, production, processing, and marketing has brought extremely high economic efficiency, it also has brought with it significant problems, even in developed countries. Indeed, in many developing countries where public institutions have little influence on environmental protection, the situation can be intolerable, when, for example, the uncontrolled use of antibiotics to maintain animal health threatens food safety. Nutrient containment and recycling from animal wastes is a serious concern. Nitrogen and phosphorus loss from manure can be a major contaminant. As agriculture intensifies, nitrate (the most troublesome soluble form of nitrogen) contamination of drinking water will become one of the most serious environmental problems. The solution to this problem will depend on the proper location of animal confinement facilities with respect to water sources. Higher locations provide greater nutrient management options. The huge amounts of feed and water used in these operations create a

TABLE 3   Comparison of the Biophysical, Social, and Economic Attributes of Land–Use Systems in the Humid Tropics

	Biophysical Attributes				
	Nutrient Cycling Capacity	Soil and Water Conservation Capacity	Stability Toward Pests and Disease	Bio-diversity Level	Carbon Storage
Intensive cropping					
Low-resource areas	M(H)	M	L(M)	L	L
High-resource areas	L(M)	L(M)	L(M)	L(M)	L
Low–intensity shifting					
cultivation	L	L(M)	L	M	L(M)
Agropastoral systems	M	M(H)	M	M	M
Cattle ranching	L	L(M)	M	L(M)	L(M)
Agroforestry	M	M	M	L(M)	M
Mixed tree systems	M	M(H)	M	M	M(H)
Perennial tree crop					
plantations	M	M(H)	L	L	M
Plantation forestry	M	M(H)	L(M)	L	M(H)
Regenerating and					
secondary forests	M	M(H)	M	M	M(H)
Natural forest					
management	M	H	M	M	M(H)
Modified forests	M	H	M	M	H
Forest reserves	M	H	H	H	H

NOTE: L (low), M (moderate), and H (high) refer to the level at which a given land would reflect a given attribute, using widely available technologies for each land use system. Symbols in parentheses indicate potential of best technologies assuming continued short–term (5- to 10-year period) research and extension.

SOURCE: National Research Council, *Sustainable Agriculture and the Environment in the Humid Tropics* (Washington, D.C.: National Academy Press, 1993), 140–141.

very large volume and tonnage of nutrient and bacteria-rich water. A facility located in the upper part of the landscape can distribute much of the waste by gravity to the agricultural land where it has great benefit. A poorly situated facility must undertake expensive containment, pumping, and long-distance hauling operations. In short, few developing country animal confinement facilities are now sustainable. Moreover, it is highly probable that regulatory control will be essential for long-term public acceptance. Fly and odor problems, which are related and dependent on location, require specific attention. Livestock operations in Japan are facing the enormous social and environmental pressures from flies, odor, and waste disposal soon to be felt in most countries, where agriculture coexists with an increasingly intensive and diverse use of lands.

Modern industries are just beginning to evolve around many nonfood agricultural products, many of them included in a broad range of crop, livestock, and tree-producing systems that are better suited to lands unfit for cereal crops. Specialty wood such as rattan for furniture, bamboo for construction, palm fronds for

Social Attributes			Economic Attributes		
Health and Nutritional Benefits	Cultural and Communal Viability	Political Acceptability	Required External Inputs	Employment Per Land Unit	Income
M(H)	H	H	H	H	H
M(H)	M(H)	M(H)	M	M(H)	L(M)
H	M	M	M	H	M
M(H)	M(H)	M(H)	M	M(H)	M
L(M)	L(M)	H	M	L	M
M	M(H)	M(H)	M	L(M)	L(M)
L(M)	M(H)	M(H)	L(M)	L(M)	L(M)
L	M	H	H	M	H
L	M	H	M(H)	M	M
L	M	M	L	L	L
L	M(H)	M(H)	M	M	M
L(M)	M	M	L(M)	M	L
L	L	M	L	L	L

basketry and handicrafts, medicinal herbs, wood for charcoal, fodder trees for animal feed, landscape nursery plants, and even fresh market cut flowers are derived from farming systems that have intricate combinations of annual and perennial crops and a variety of animals. Farm output is used for family consumption and for sale to local markets. Many high-value products from such systems—such as handicrafts, spices, and a large number of animal products—even reach national and global markets as national economies develop.

These production systems have a wide range of biological, social, economic, and environmental impacts (Table 3). Many have a high level of environmental sustainability on fragile soils—that is, they have high nutrient recycling and retention and high biological stability, thereby requiring few if any pesticides—and many are labor-intensive, with a quite high economic return on labor. They are usually small scale, are well adapted to small land holdings, and are reasonably specific in their environmental, economic, and sociopolitical applications. Finally, they usually produce small amounts of each product and so are adaptable to specific market opportunities. Because many plant species grow only in certain environments, there is no fixed formula or pattern for how the thousands of plant and animal species should be assembled in a "pre-designed" system.

Any national agricultural system is composed of a mixture of such production systems (Figure 3), which today are proliferating in response to the dramati-

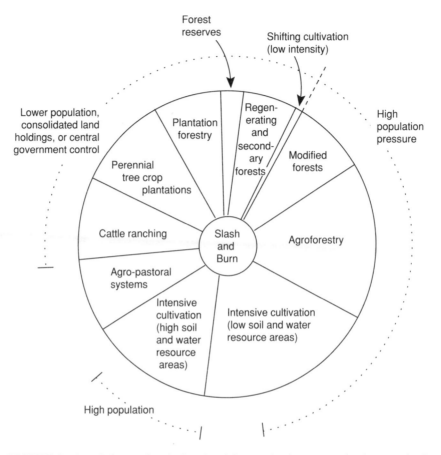

FIGURE 3 Area balance of agricultural and forestry land-use types that have evolved over past decades from native forest in a typical developing country. Land clearing is usually by slash and burn, and moves toward alternatives uses are based on soil type, land tenure, farm size, market availability, and other factors such as domestic security and availability of capital.

cally increasing markets in the industrially developing countries such as Indonesia, Thailand, and Korea. Increasing population pressure on the dwindling supply of agricultural land in these countries is forcing an evolution toward intensive, highly diverse systems at the expense of less-intensive grazing or shifting cultivation. These land-use changes are not without conflict. Population pressures in the industrialized countries cause breakdown of traditional (often tribal) land tenure arrangements. In countries such as Brazil with a large land base, the evolution from forest to slash and burn and the move toward either large-scale cattle ranching or intensive smallholder use is often fiercely contested, with political, legal, and economic powers pitted against smallholder interest.

To meet a range of market and social needs and to operate within local environmental and resource constraints, farm structures in developing countries must take several forms. Industrial enterprises, operated under appropriate social and environmental controls, will play an important role, as will the limited-diversity, mostly small farms in the high soil and water resource areas, which will continue to serve as their countries' breadbaskets. Appropriate systems for the increasingly populous, environmentally sensitive areas that are now under great stress will have high biological diversity and a high proportion of perennial plant species and animals, and will be highly integrated both biologically and socially. Because of the requirements for perennial species, such systems must maintain greater amounts of carbon in their plant and animal biomass than do systems in high soil and water areas. Carbon accumulates in the system as a result of photosynthesis and plant growth, being "fixed" from carbon dioxide in the air as part of trees, roots, organic matter in the soil, and in plant and animal residues. In fact, these systems, at least during the preindustrial stages of economic development, would operate very much as a carbon-based economy; carbon accumulated in its many diverse forms in the system would act as "biological capital"—much like the financial capital that must be accrued by other businesses to prosper. Economic and social conditions must favor a long-term outlook for farmers to permit carbon stocks to accumulate to levels that form a base for nutrient recycling and high productivity.

SOLUTIONS TO KEY PRODUCTION PROBLEMS

The technological breakthroughs and the industrial systems emerging in agriculture share many real but solvable problems, illustrated by a few of the more pressing examples, in addition to those mentioned earlier.

Single-commodity, "industrial" crops are associated with heavy pesticide use, soil erosion, and waste disposal problems (processing waste—plastic materials in the case of bananas). The solutions to these problems—integrated pest management, some degree of landscape diversity, and careful soil management—are complex and often costly. Thus regulation and subsidies or incentives probably will have to be a driving force behind the adoption of such solutions. As large-scale, vertically structured industries come under increasing international and local regulatory pressure, they will develop and employ the corrective technologies, but production costs will increase. Publicly funded research or correction should not be necessary in most instances.

Intensive, small-scale systems in high soil and water resource areas fall prey to problems stemming from pest management (pesticide loading), crop nutrient availability and loss, soil erosion, and salinity (rising water tables). This area will require considerable public technical assistance through traditional public research and extension channels. The breeding of pest- and disease-resistant crops, predator management, and farmer training in integrated pest management (IPM) methods are crucial to reducing the need for pesticides.

Nutrient containment and recycling will become increasingly important as systems continue to intensify and as nutrient inputs are increased to raise productivity. Work in this area has focused on the economics of higher efficiency, but as water supplies become increasingly threatened by high nitrate levels, public pressure to halt such contamination will increase. Current work in the management of soil biota for nutrient mobilization and recycling (containment) within soil shows considerable promise. Such processes are common to the intensive, mixed-culture systems of traditional agriculture. The exciting news is that these very efficiencies seem to work under the high nutrient flow and turnover rates of the high production systems in developed countries. This is one of the most promising lines of research toward sustainability and cost effectiveness in highly productive systems.

Water management, in terms of both supply and drainage, often has a large public component. During the construction of water systems, budget constraints often lead to inadequate drainage systems. This results in rising water tables, thereby turning large areas into wetlands, and to the accumulation of soluble salts, which reduce the production potential of the land. Unfortunately, the measures needed to correct these problems after the damage occurs are extremely expensive and may take years.

Mixed systems for lower soil and water resource environments suffer from lack of land tenure permitting long-term investment, soil erosion and lack of nutrients, as well as lack of infrastructure, social stability, and security in the community. In fact, problems in areas having these systems are as much social and political as they are technical. These areas often are newly settled with no land tenure history as in the better soil areas. Social and political institutions are weak, frequently with little security. The productivity of these areas depends on a mix of perennial crops, livestock, and land improvement, all of which require adequate conditions for long-term investment. A farmer's investment in long-term crops (the accrual of biological capital) will require more labor than cash but will require the same investment environment as will the control of soil erosion.

Ultimately, the cycling of nutrients becomes limiting in such systems. Thus eventually the ecological processes for containing and recycling applied nutrients must be mastered so that such mixed systems can be sustained. These systems are, after all, critical to the support of rural populations, to national productivity, and to overall political and social stability.

In addressing the problems of how to use resources efficiently and how to maintain the environmental stability of the many highly promising technologies and emerging industrial systems, scientists have realized that they know little about the ecological processes involved. Enormous potential exists for harnessing soil biota to more effectively release soil nutrients at low levels of availability[10] and to contain nutrients at high flow rates by achieving greater synchrony between soil nutrient release and crop uptake.[11] Likewise, the potential for enhanced pest and disease management by manipulating ecological processes has barely been tapped. But even though more effective harnessing of these processes will improve the efficiency of nutrient recycling, it will not eliminate the need for fertilizers. Similarly, more effective pest management will reduce but not eliminate the need for pesticides. Ecological methods thus hold great promise for increasing production efficiency and reducing the environmental damage caused by high-productivity agriculture, but the development of the scientific basis for production systems ecology will require a major input from the public sector.

PATHWAYS OF CHANGE FOR DEVELOPING COUNTRY SYSTEMS

Three basic, often overlapping pathways for the generation and movement of capital, technology, knowledge, and production resources are commonly found throughout agriculture. Each pathway has a range of options, depending on the stage of agricultural development and the specific circumstances encountered, but the three main pathways are each suited to particular agricultural situations. Most rural development uses combinations of the three.

The private sector industrial-dominant pathway is the most narrowly focused and specific. In this model, technology is highly privatized and, in many cases, may be proprietary. The industries tend to be vertically integrated economically and capital-intensive. This model is most applicable where standardization and control of the production process are possible, where local environmental and social interaction are modest or low (or are disregarded), or where a homogeneous environment exists. Large-scale confinement animal operations and plantation crops for export such as sugar and oil palm are examples.

The public sector land-grant, research and extension-dominant model with its many variations is the second pathway. In this model, a high proportion of technology is generated within public sector institutions, with delivery to farmers heavily dependent on public channels. This pathway has a high degree of centralization of both technology and of knowledge. It is the most applicable to the small number of crop and animal commodities that constitute most agricultural value and are widely produced. Traditionally, this pathway has served well in production areas of high homogeneity and high production potential where a modest number of research and development sites are representative of broad areas. Its reach extends generally toward a greater diversity of production environments than that of the industrial pathway, but, because of its centralization, it has significant weaknesses in dealing with a high level of biological integration in very diverse production systems or with high levels of social integration. The well-known green revolution approaches are excellent examples of the effective use of this pathway.

The third pathway is a decentralized, networking-type model in which the energies for change are concentrated at the community-based social and political levels, with a much greater proportion of technology and information flow being horizontal. (Pathways for change that operate with a high degree of farmer-to-farmer linkages in no way imply communal ownership of land.) The decision-making process for the selection and application of technologies is more decentralized than in the first two pathways, with technologies and resources entering the system at widely scattered points. Thus the system requires a higher level of social infrastructure and networking than do the more centralized pathways. In such systems, the portion of indigenous, or community-based, knowledge is high in relation to new knowledge or technology entering from outside. Outside sources are usually needed for solving specific problems or understanding a process.

These three pathways for agricultural development are systems-specific. Nevertheless, the proportionate mix of each pathway must be carefully determined for each type of agricultural system (each agricultural sector).

Development of the industrial sector is highly dependent on the investment climate, which, among other things, depends on social and political stability and on government policy. Although there is considerable international experience in this area, the environmental and social acceptability of such operations, given the relatively undeveloped regulatory frameworks existing in most developing countries, is sometimes abysmal. Some multinational companies, such as the Cargill poultry processing plant in Thailand, have done a good job in treating and recycling wastewater, but, overall, industry's record is not good. As a result, the developing countries, like many of the more developed countries, will undertake greater regulation to correct the most flagrant abuses. Generally, the creation of new technologies through research and development for these industries will continue to be in the private sector, and their spread will be through market channels.

The high soil and water resource areas will continue to depend heavily on public sector international and national research programs for their genetic resources and nutrient and integrated pest management technologies. Recently, however, this international research sector has been under enormous financial constraints, brought about in part by the global recession, by alternative demands for resources, and by a global complacency about food sufficiency for the next decade or more.[12] Within the high soil and water resource areas, considerably more location- and time-specific practices must be used to achieve more efficient pest and nutrient management and to reduce environmental loading. This can be accomplished by adding a much broader base of local area networking and farmer-to-farmer approaches to the more traditional public sector pathway. In addition, the technological inputs from private sector agribusiness must continue to grow.

The highly integrated systems of the lower productivity areas present a much greater challenge, and public sector institutions, especially the International Centre for Research in Agroforestry (ICRAF), have been reaching out to them.[13] One farmer-to-farmer, local community initiative in agroforestry has been described as "a new path." ICRAF's Southeast Asia program is documenting the application of three types of agroforestry systems to serve as a basis for local adaptation by farmers.[14] A technology flow pathway based on farmer and community networking and empowerment is crucial to these highly integrated systems. Market forces as well as pricing policy must be favorable.

In contrast, scientific understanding of the highly diverse, very location-specific farming systems is extremely difficult. Their improvement without such knowledge is nearly impossible. In a farmer-participant research model now "fashionably" popular with scientists, farmers are considered members of the research team. They operate the system and work with scientists to collect and interpret descriptive data. This approach has had mixed success in both understanding the systems and changing them. Thus the value of this approach, as it has been practiced for the last 20 years, must be questioned. There have been few common themes and virtually no hypotheses suggested for systems structure or function, and the vast array of systems descriptions has not generated a body of knowledge useful to either researchers or farmers. The ICRAF Southeast Asia approach, based on hypotheses of systems structure, is exemplary in its innovation.[15]

Any effective strategy for the agricultural development of the low soil and water resource areas must seek the participation of the public sector in (1) creating an enabling environment—public security, guaranteed access to land, and appropriate pricing and economic incentives; (2) assisting in the development of a progressive rural infrastructure—private rural institutions, a framework for farmer interaction, and a modest physical infrastructure; (3) evolving indigenous lead technologies in land and soil management, efficient use of water, and carbon husbandry; and (4) providing new technologies in such areas as tree production, animal husbandry, and high-value crops for market.

CONCLUSION

The greening of agriculture will require that its structure continue to evolve, with strong guidance and help from public policy and investment. Appropriate resources, incentives, and controls must be applied not to a single pathway but to the three overlapping development pathways. Far more emphasis must be given to the empowerment of local community groups, particularly in areas of complex and highly integrated farming systems. In so doing, their access to the scientific community must be strengthened. If indigenous knowledge alone were able to evolve quickly enough to meet changing social, economic, and environmental demands, the developing countries would not be facing such overwhelming problems. But at the same time, top-down solutions to these problems are illusory. Research needs will continue to increase for the near term, although, unfortunately, the public's interest in research is waning.

The technological breakthroughs and the adjustment of rapidly evolving production systems for greater sustainability will continue to require major public sector support, much of which will have to come from multilateral agencies.

The agricultural development future, with its plethora of options and the increasingly urgent demands of a rapidly growing population, is far more complex than it was during earlier decades of the green revolution. But in spite of the challenges, there are many reasons for optimism.

NOTES

1. Rockefeller Foundation, *1993 Annual Report* (New York: Rockefeller Foundation, 1993), 21.

2. P. Pinstrup-Anderson and R. Pandya-Lorch, "Alleviating Poverty, Intensifying Agriculture, and Effectively Managing National Resources," International Food Policy Research Institute, Washington, D.C., 1994, 7.

3. P. Crosson and J. R. Anderson, "Resources and Global Food Prospects: Supply and Demand for Cereals to 2030," World Bank Technical Paper No. 184, World Bank, Washington, D.C., 1992.

4. R. Lal, G. F. Hall, and F. P. Miller, "Soil and Degradation: I. Basic Processes," *Journal of Land Degradation and Rehabilitation* 1 (1989): 5-69. To further explain what is shown in Figure 2, the productivity of agricultural soils is dependent on a layer of geologically-formed topsoil, which may range in depth from a few inches to several feet. Degradation of the topsoil can include loss resulting from wind, water movement, or landslide; compaction from heavy machinery use, particularly when the soil is wet; loss of plant nutrients and organic matter; or buildup of salt or other unwanted toxins. A resistant soil is often thick, flat in topography—and thus less subject to erosion—and rich in nutrients, allowing it to tolerate heavier and lengthier stress. Once such a soil is severely degraded, its agricultural productivity is nearly zero, like that reached sooner by marginal soils that have been abused.

5. These leaders successfully set forth a development paradigm that was adopted rapidly in a climate of extreme urgency stemming from a perceived imminent global food scarcity. See A. Mosher, *Getting Agriculture Moving: Essentials for Development and Modernization* (New York: Praeger, for the Agriculture Development Council, 1966).

6. United Nations, *Agenda 21,* Report of the United Nations Conference on Environment and Development, Rio de Janeiro (New York: United Nations, 1992).

7. Consultative Group on International Agricultural Research, *A CGIAR Response to UNCED Agenda 21 Recommendations* (Washington, D.C.: World Bank, 1992).

8. D. J. Greenland et al., "Soil, Water, and Nutrient Management Research—A New Agenda," International Board for Soil Research and Management, Bangkok, Thailand, 1994.

9. D. P. Garrity, "Agroforestry: Getting Smallholders Involved in Reforestation," International Centre for Research in Agroforestry, Bogor, Indonesia, 1994.

10. P. L. Woomer and M. J. Swift, eds., *The Biological Management of Tropical Soil Fertility* (Exeter, England: Tropical Soil Biology and Fertility Programme and Sayce Publishing, 1994).

11. R. R. Harwood, "Managing the Living Soil for Human Well-Being," in *Reinventing Agriculture and Rural Development*, ed. S. A. Breth (Morrilton, Ark.: Winrock International, 1994).

12. R. O. Blake et al., "Feeding 10 Billion People in 2050: The Key Role of the CGIAR's International Agricultural Research Centers," Action Group on Food Security, Washington, D.C., 1994; Consultative Group on International Agricultural Research, *A CGIAR Response*; M. C. Ageaoili and M. W. Rosegrant, "World Supply and Demand Projections for Cereals, 2020," International Food Policy Research Institute, Washington, D.C., 1994; M. C. Ageaoili and M. W. Rosegrant, "World Production of Cereals, 1966-1990," International Food Policy Research Institute, Washington, D.C., 1994; and Pinstrup-Anderson and Pandya-Lorch, "Alleviating Poverty."

13. International Centre for Research in Agroforestry, "Agroforestry for Improved Land Use: ICRAF's Medium-Term Plan, 1994-1998," ICRAF, Nairobi, Kenya, 1993.

14. D. P. Garrity, "ICRAF Southeast Asia: Implementing the Vision," International Centre for Research in Agroforestry," Bogor, Indonesia, 1994.

15. Ibid.

Lessons from the Evolution of Electronics Manufacturing Technologies

SIDNEY F. HEATH III
Technology Planning Director, AT&T

The global electronics market, which grew from $220 billion in 1980 to $654 billion in 1990, will reach an estimated $1.405 trillion in 2000.[1] Because of this tremendous growth in demand, electronics manufacturing is one of the main sources of advances in technology and innovation. Worldwide, electronics manufacturing productivity has been a primary factor in sustaining and increasing national standards of living and economic growth by adding value to raw materials. Indeed, electronic manufacturing's economic pump, when it can generate growth in productivity, delivers a far larger percentage of added value to the economy than that of any other single sector. These benefits also are available to developing countries that are able to successfully nurture an electronics manufacturing sector. But to do so, these countries need to understand the forces driving the rapid technological changes occurring in this sector and the basics of the manufacturing technology shifts.

FORCES AFFECTING ELECTRONICS
MANUFACTURING TECHNOLOGY

Three main forces are driving innovation in electronics manufacturing technology: information technology, globalization, and intensified competition.

The major trends in information technologies—electronics photonics, speech process, video, computing, telecommunications, software, and terminals—are described in "Information Technology for Development" by John S. Mayo. Powerful technological trends in silicon, photonics, and software development have produced the technologies that are supporting the emerging multimedia commu-

161

nication revolution and the evolution of a national information superhighway. For example, the silicon components at the core of all electronics devices have been doubling in speed every two years; compression algorithms have allowed technologists to encode the same information in one-thirtieth the space every five years; lasers are doubling in speed every three to four years; data storage costs are being cut in half every other year; and software has generated a computer literacy requirement that affects most occupations. Overall, these trends will determine what can be manufactured and how manufacturing will operate. Most electronic assembly operations will use similar components, interconnection schemes, packaging, and power supplies.

The need for reliable miniaturization processes in electronics manufacturing has driven such operations into suprahuman activities that require automation support to deliver a product that meets global quality and reliability standards. This in turn has meant a reduced role for unskilled labor and a greater dependence on a highly skilled work force.

Effective manufacturing operations also require the seamless integration of the manufacturing technologies into a full-stream, high-velocity process. Individual technology steps and their facilities are readily available in the market. Thus full-stream integration and effective operation require either significant investments in technology transfer or a long, expensive internal development interval. Fortunately, as the cost of processing MIPS (million instructions per second) has dropped, the cost of computer-integrated manufacturing (CIM) has dropped as well and provided for effective process control and concurrent engineering.

Globalization, the second force affecting innovations in manufacturing technology, has changed the context for manufacturing decisions in electronic assembly. Manufacturing operations can no longer exist effectively as stand-alone entities because they must depend heavily on their global supply base for components, manufacturing infrastructure, and the development of manufacturing processes. In the developing countries, policies and regulations on trade policies, soft financing, economic offset, and duties and taxes on the flow of these manufacturing elements can significantly affect the climate for manufacturing. Furthermore, the competitive complexion of a manufacturing entity is becoming increasingly dependent on the successful utilization and leverage of its global production network. Without the support of the developing countries themselves, the manufacturing operations will not be globally competitive.

Finally, the globalization of manufacturing has intensified the competitiveness to be first to market, to be customer-responsive, and to have the highest productivity rates. Being first to market generally means higher profit margins and larger market shares. But such a focus on time-based processes requires the adoption of quality principles and total quality management. Thus to be globally competitive, manufacturing companies must master the basics of continuous learning and improving through a quality orientation. They then can take the next step, which is to drive time from their processes.

Given the pace of technological change and the increasing competitiveness of the global marketplace, rates of productivity improvement are a critical determinant of financial success. Over the long term, successful manufacturing companies can realize productivity rates twice as high as those of their competitors or even higher.[2] Indeed, operations that focus on productivity, create an innovative environment, and drive for results can actually outperform average operations two to one. This capability then becomes a long-term differentiator that is difficult to duplicate because the requisite skills have been deeply imbedded in the work force. Furthermore, because the advantages of proprietary technologies have been disappearing, and the returns on sales, revenue growth, and product life cycles have been declining at double-digit rates,[3] successful global manufacturing operations must leverage their learning and integration across their entire production network where capabilities and learning are done globally. This in turn implies that each location will depend heavily on the skills and resources at other locations. Developing countries that can establish a climate conducive to building systematic, comprehensive approaches to continuous innovation and productivity will find that manufacturing sectors are better able to thrive.

TRENDS IN MANUFACTURING TECHNOLOGY

Any understanding of the shifts in electronics manufacturing requires an in-depth look at the trends affecting the five major manufacturing processes: manufacturing assembly, designer interfaces, supplier interfaces, order realization, and distribution. This paper, however, will focus only on the first three processes, as well as the people who carry them out (Figure 1).

People

The organizational structure of manufacturing and people's roles and responsibilities in that structure are evolving from an environment geared toward managing people to one directed at coaching people and managing processes. Leading-edge manufacturers have been eliminating layers of management and developing effective communication capabilities for flatter organizations. Management is no longer in the loop for all decisions; in its coaching role it now primarily sets direction, establishes commitment, and marshals resources.

In the 1970s and early 1980s, employees at all levels were given relatively narrow job descriptions for specific, well-defined tasks because it was assumed that processes would remain stable. But in the 1980s, the realization dawned that processes evolve and therefore so should tasks. This notion called for pushing decision making down to the lowest possible level where the employees could add value by improving the process. Indeed, differential advantages could be obtained through empowered production associates who could identify and solve problems.

164

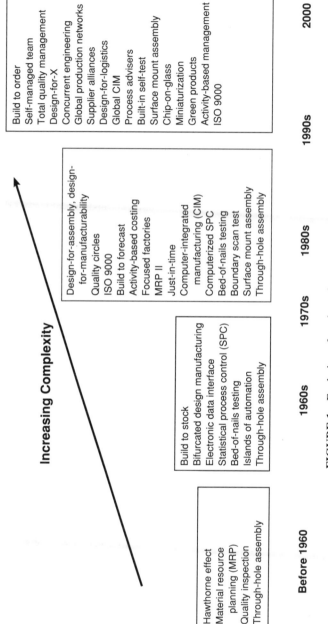

FIGURE 1 Evolution of modern electronics manufacturing.

This is one of the most misunderstood issues when evaluating the transfer of manufacturing technology. Facilities and buildings can be established in a relatively short period of time, but the development of the human capital is fairly invisible, expensive, difficult to assess, and the most time-consuming—yet it is by far the most important aspect of technology transfer. Government policies that do not encourage a long-term bond between employer and employee further complicate this process. For example, if retirement programs are maintained by the government, employees are more apt to transfer to new employers after they have been trained by their present one.

Where self-directed teams are implemented, improvements in all areas of manufacturing—such as quality, productivity, cycle time, worker safety, and customer satisfaction—are realized. In addition, layers of bureaucracy and many manufacturing costs can be eliminated. Yet, perhaps more important, workers are better able to adapt to the fast pace of technology change, thereby producing a more agile organization seeking to continually improve its ability to thrive in a competitive environment of continuous and unanticipated change while nimbly responding to rapidly changing, customer-driven markets. This process of leveraging intellectual capital is key to the new manufacturing paradigms because this is where the wealth of the manufacturing competencies resides, not in the manufacturing facilities. When this process is carried out properly, performance may improve by an order of magnitude over that of companies not at the leading edge.[4] It must be remembered, however, that the effective utilization of high-performance teams will erode the advantages of low-cost labor unless the latter are similarly trained and empowered.

When high-performance teams are in place, manufacturing management will have the time and energy to tackle the kinds of important strategic and financial issues facing any operation seeking to continually improve its operations. But what kind of management can provide the leadership and direction required? They are manufacturing professionals who are well versed in finance and who have social skills in addition to their knowledge of manufacturing. Fledgling manufacturing operations in developing countries may rely on costly expatriates to provide these qualities, but over the long run this is no substitute for the development and training of national manufacturing managers in order to be competitive and to harmonize the manufacturing operation to local environments. Such development and training will require the longest lead time of all the developmental processes.

Manufacturing Assembly

In the early 1920s, *manufacturing quality* consisted primarily of product inspection for conformance. Statistical process control (SPC) was introduced in the 1940s to separate out assignable causes from natural variation and to monitor processes. But in the 40 years that followed, the management approach to SPC was misplaced, with the exception of Japan. In the mid-1980s, the responsibility for product quality was shifted from a staff organization of a few to everyone, the basis of total quality management (TQM). Juran's philosophy that most quality problems are related to the management systems and Deming's 14 points of quality management revitalized quality through the basics; such simple problem-solving tools as pareto charts, histograms, and fishbone charts produced amazing results. Quality became a strategic issue—so much so that many companies documented their processes and became ISO 9000 registered in the late 1980s.

Today, this emphasis on quality is extending into programs that provide customer feedback on the competitors' performances. In fact, customer surveys have become a key source of customer data for directing quality improvements. Customer value-added (CVA) analysis (a way of looking at oneself through the eyes of one's customers) further refines customer feedback and serves as a basis for proactive improvement programs. One important aspect of the customer focus is the notion of competing through time and understanding the entire customer relationship.

In the early 1990s, the availability of personal computers allowed SPC calculations to be figured in real time, cost-effectively. Many factories have continuous automated data collection and process control chart monitoring, allowing the ongoing analysis of feedback data. As trends are discovered and analyzed, processes are optimized. Two major developments in this area are, first, a move toward designing a process right the first time and including sufficient robust parameters so that data collection and charting are not necessary, and, second, the implementation of new hearty data collection systems that will accommodate product flexibility and early ramp-up of new designs.

During the last five years, the science of environmental stress testing has matured, improving the reliability of the product as seen by the customer. This was carried out first in the manufacturing process and has now moved upstream into the concurrent design process during the early prototype stage. In the current environmental stress testing, new designs are functionally tested to establish upper and lower temperature bounds with voltage variations. Emerging technologies include clock variations, faster temperature fluctuations, shock, mechanical vibrations, electrostatic discharge (ESD), and artificial lightning.

The next phase of process control is off-line real-time video monitoring of manufacturing processes to strengthen the understanding of the underlying manufacturing science. The thrust is beyond six sigma expectations (quality levels) for manufacturing operations to six sigma goals for entire manufacturing lines. Leading-edge manufacturing operations will have performance levels that qualify for recognition by government and international quality awards—for example, the Deming Prize, European Quality Award, and Baldrige Award. Their manufacturing operations will perform at these levels regardless of their country of operation.

Customer requirements and competitive pressures will demand a manufacturing management discipline that concentrates on continuous improvement in product and process quality. Without this passion, the manufacturing operations will not be able to compete globally. Developing countries can foster this climate by offering quality recognition awards that require the very highest levels of quality excellence, encourage quality training, and sponsor quality registration programs.

In *packaging,* electronics manufacturing technology is evolving quickly to keep pace with the silicon density revolution. The trend toward finer pitch and denser electronic assemblies has enabled moves toward densification to provide more functionality in the same space and miniaturization to produce a product that is smaller than similar products of the recent past.

During the early 1980s, electronic assembly focused primarily on leaded devices that were placed in holes on printed wiring boards (PWBs) and then wave soldered. In the mid-1980s, surface mount began replacing through-hole assembly; the main advantage of surface mount: it enables double-sided assembly and more interconnect and thus higher density. It also uses the stencil print, place, and reflow process, which is more consistent and allows finer pitch than through-hole. In the late 1980s, surface mount pitch was reduced from 0.100 inch to 0.050 inch and in the 1990s to 0.025 inch and 0.020 inch. The trend continues to push the limits further to 0.015 inch and 0.012 inch, but probably will not extend beyond without additional technology. As the pitch becomes finer, the leads are more susceptible to damage during handling and the coplanarity of PWBs affects the soldering consistency. Because the trend to surface mount designs is well under way, it is now becoming difficult to procure the older through-hole packages. This trend has contributed to a trend toward hiring fewer, but more highly skilled, employees (Figure 2).

Another package technology, the ball grid array (BGA), does not have a peripherally leaded interconnect; the input/output (I/O) are distributed over the area of the package face in an array. The main advantages of this approach are that the fragile leads are replaced with solder balls and the pitch for the same I/O is greater, thus making device to PWB assembly easier. A major benefit is that the existing surface mount assembly equipment is reusable.

The next industry step toward smaller packages, called direct chip attach (DCA), is to eliminate the package altogether and attach the silicon chip directly to the substrate. The method of attaching the chip directly with conductive adhesive, electrically interconnected with wirebonding, is called chip-on-board. An important approach, known as chip-scale packaging, puts an interlayer on the chip for protection during handling and also provides compliance. When the chip is placed on glass substrate (as in displays), the arrangement is called chip-on-glass. When the chip is placed face down, directly on the substrate or glass, it is called flip-chip. In the cases of direct chip attach, the packages are more susceptible to damage from electrostatic discharge and require other than conventional surface mount assembly equipment.

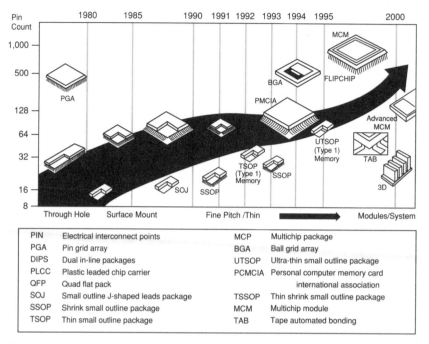

PIN	Electrical interconnect points	MCP	Multichip package
PGA	Pin grid array	BGA	Ball grid array
DIPS	Dual in-line packages	UTSOP	Ultra-thin small outline package
PLCC	Plastic leaded chip carrier	PCMCIA	Personal computer memory card
QFP	Quad flat pack		international association
SOJ	Small outline J-shaped leads package	TSSOP	Thin shrink small outline package
SSOP	Shrink small outline package	MCM	Multichip module
TSOP	Thin small outline package	TAB	Tape automated bonding

FIGURE 2   The evolution of packaging.

Traditionally, electronic products have been tested, first, for structural integrity (presence of parts, orientation, continuity) and then for function. The structural test used a "bed of nails" and sophisticated test computers to test continuity on many circuit paths through small probes ("nails"). Functional testing checked the overall PWB function through its I/O connector(s). As packaging technology leaned toward finer pitch and more functionality in less space, space for the bed-of-nails testing became limited. Probe pitch in the 1980s was typically 0.100 inch. It improved to 0.050 inch and in some cases to 0.025 inch, but most circuits were much denser and probe points on circuits were using space on the PWBs that components could use. Thus test coverage was dramatically limited.

The new methods are the boundary scan, to overcome probe density limitations in the structural test, and the built-in self-test (BIST) for chip-level functional testing of digital devices. Use of BIST requires that a portion of the silicon chip be set aside for self-test—the test probes are etched in silicon when the device is made, and several I/O are reserved for initiating and accessing the result of the self-test. This technology is being implemented first in processors and memory devices. The boundary scan method checks the interconnect between devices. The combination of boundary scan and BIST may someday replace bed-of-nails and functional testing for many products and enable high reliability of even more complex electronic devices.

The basic interconnect material between components and the PWB is tin-lead solder, used in various proportions to achieve different melting temperatures for a variety of applications. In the 1980s and early 1990s, much progress was made toward developing water-soluble fluxes and lead-free solders, and the future will see less use of hazardous materials such as lead and some fluxes, leading to "greener" manufacturing processes.

Computer-integrated manufacturing (CIM) has progressed to the point where most manufacturing lines producing a diversity of electronic products use automatic product identifiers and download programs for assembly sequencing and process recipes. Operations without this capability have a much higher setup time and cost and cannot compete as effectively. The integration of multiple software systems allows the seamless electronic transfer of manufacturing information all the way from schematic design to building the materials lists for the material resource planning (MRP) systems. Capabilities now exist for global CIM so that designers and manufacturing locations can be effectively linked independent of their country of origin. Electronic data transfer systems can be expensive, however, if facilities have to be installed specifically for CIM data transfers. In these cases, speed and response time requirements will typically determine the communication mode.

Generally, most of the individual facilities required for manufacturing steps are available globally to all manufacturers. Support for such facilities, however, may require the availability of trained technicians on the premises and experts on call for emergency support. Technical information is readily available at trade shows and training centers and through various associations such as the Society of Manufacturing Engineers, Surface Mount Equipment Manufacturers Association, Institute of Electrical and Electronics Engineers, and NEPCON electronics exhibitions. Significant improvements in processes become well known and available relatively quickly, yet the performance of a manufacturing operation is driven primarily by the effective integration of many processes and operating procedures, for which the integration capabilities or recipes generally are not readily available and, in reality, are dependent on the competencies in the manufacturing operation. Access to these capabilities or recipes is best obtained through technology transfer agreements such as joint ventures or licensing. Without an effective technology transfer program, developing countries can easily choose the wrong equipment (which may be right for other applications) or not integrate the equipment correctly.

Design Interfaces

In a competitive environment that is global, intense, and dynamic, the development of new products and processes is becoming the focal point of competition. Manufacturing companies that can get to market faster and more efficiently with products that are well matched to the needs and expectations of the customer create significant market leverage in both market share and margin.

In the 1960s, design and manufacturing had separate departments of specialists. Designers converted marketing's features requirements into a set of drawings and other documentation that specified a product for manufacturing to build. Design was further subdivided into process and product design. Working within these organizational dividers ("walls"), marketing threw requests over the wall to product design; in turn, process designers and product designers threw their specifications over the wall to manufacturing. As a result, manufacturing built products without a deep understanding of the customers' needs.

In the 1970s, it was recognized that if designers had a better understanding of how a product is built, they would design products that were easier and cheaper to manufacture. Soon, then, design-for-assembly (DFA) analysis tools were developed and applied. These tools stressed the importance of minimizing part counts by expanding the functionality of design elements and minimizing handling and orientation changes. From the DFA push evolved a more general look at design-for-manufacturability (DFM) and then design-for-X (DFX), where X has come to mean any downstream process such as design-for-logistics, design-for-test, design-for-installation, or design-for-environment.

In the electronics industry, the increase in silicon density and consequently the ability to provide more features and functionality in the same space also have contributed to closer ties between manufacturing and design. Competition to provide advanced consumer electronics products to the marketplace has led companies to capitalize on the advantages of closer manufacturing and design organizations and the breaking down of the former walls. Design has evolved from a serial-design/build/test work mode to a model-and-simulate/test-build/verify concurrent engineering approach.

Trends in information technology have decreased the costs of design workstation platforms by 10 percent a year and increased their computing power by 30 percent a year. The development of information systems has become more efficient because of database independence foundations and object-oriented design methodologies. Global movements of product design data, facilitated by improvements in local area and wide area networks, have enabled designs to be simulated at the circuit pack and system level before any prototypes are built. Although the capability to use simulation of firmware and diagnostic software exists, cost-effective computing power is still a bottleneck for the aggressive projects.

Many of the DFX rules are built into the CAD (computer-aided design) tools. Now cross-functional teams address and overcome issues in the early stages of a new product, allowing manufacturing to participate early in the product development cycle and resulting in more manufacturable, quicker-to-market, cost-effective products. Information technology has provided the capability to elec-

tronically carry out concurrent design and reviews at different geographic locations. Because many electronic product life cycles are now measured in months
instead of years, the initial design must be correct the first time. There is no
longer time to cycle through iterations of design.

Since as much as 75-90 percent of manufacturing costs are determined in the
design stage, effective concurrent design in product and process is paramount to
bringing new products with superior performance to market. Often the design
architectures are determined by global technology trends (for example, micropro-
cessors or memories) that basically determine how costs are distributed between
materials and labor and load (overhead) costs. Even the options between different
manufacturing operations (such as providing for more or less labor) are becoming
more limited. The percentage of manufacturing costs assigned to material has
stayed relatively flat over the last 10 years for any particular product family, but
more products are like consumer products, with a lower percentage of labor and
load. In addition, the labor and load content has shifted from the workers directly
assembling the product to the support personnel, and this shift is particularly
acute in small factories.

For developing countries, these trends make the transition to manufacturing
more challenging and reduce the global advantage of having low-cost labor in an
unskilled work force. These challenges can be addressed by fostering a climate in
which local manufacturing can leverage off a global production network for the
initial development and shake-out of new products and processes. The leveraging
requires effective global communication networks, efficient import/export policies, unrestricted travel, joint ventures with greater than 50 percent ownership by
foreigners, and localization expectations consistent with realistic economic assessments. The developing countries also can foster the development of a skilled
work force, well trained in the quality basics along with the basic sciences.
Sponsorships of trade shows, professional associations, and training centers also
will foster the right learning environment.

Supplier Interfaces

Because 75-90 percent of the value of typical electronic products is determined by the suppliers of the components, the relationship between component
suppliers and the assembly operations has changed significantly over the last 10-
15 years. Material resources planning (MRP) enabled organizations to plan and
execute production and material requirements effectively in complex environments. It thereby introduced discipline and structure to the relationship between
the manufacturing operation and the supplier. In the mid-1980s, just-in-time
(JIT) manufacturing focused attention on reductions in waste and lead times as
well as throughput improvements. In the push to minimize inventory, suppliers
were asked to ship in smaller and smaller lot sizes with higher frequency. This
shift to just-in-time inventory drove suppliers and producers closer together.

Thus relationships with suppliers have evolved from arm's-length customer-supplier ones to long-term, almost coproducer ones. Multiple suppliers no longer compete on price for the same part; now the best suppliers are sought out and retained as a primary source that will provide the lowest world-class total cost of ownership. Price continues to be important, but other criteria (delivery, quality, proximity) are given appropriate consideration.

In this closer relationship, which allows producer and supplier to work together, producers help to improve supplier quality and suppliers participate in plans for new designs. As a result, component defect levels have dropped drastically. Multiple-level, two-way contacts allow broad sharing of process, quality, and schedule information. To facilitate the process improvements, many manufacturers now have formal quality review procedures in place. To facilitate sharing of information, corporations with larger stakes in each other's mutual success have linked key material databases and synchronized schedules.

Producers and suppliers are typically located in geographical clusters to facilitate JIT. As the distances between suppliers and assembly decrease, so does the inventory as a percentage of sales. Leading-edge procurement (LEP) arrangements, techniques, or processes are established to streamline transactions, eliminate purchase orders or manual intervention, and add valued services. As a result of such steps, suppliers are able to provide their clients with the benefits of inventory reduction, supply assurance, and on-site support through such processes as consignment of materials, dock-to-stock shipments, breadman and demand pull. LEP transactional provisioning converts the previous manual transactions into electronic data interchange using such systems as Conversant (voice messages) or EDI/EFT/EDS (electronic data interchange/electronic funds transfer/electronic data systems). This trend of supplier and manufacturer working together to eliminate wasted costs and time is likely to continue (Figure 3).

Localization of suppliers for any manufacturing operation is a significant concern when establishing operations in a developing country. Bulky, low-technology parts such as plastic and metalware are usually the first to be localized, although even this effort is often underestimated because these components often require unique properties and high quality. The value of these products will vary by product family, but the low-technology parts are typically less than 10 percent of the material cost of the final electronic product. The next level of localization requires significantly more technology transfer for such technologies as printed wiring boards and passive components. The final localization steps into the high-technology areas of integrated circuits and displays will probably be taken only if there are significant market opportunities to justify the large capital investments. Given these considerations, it is important that developing countries have an effective infrastructure and transportation system that can handle the importation of the components for assembly during these countries' initial thrust into a manufacturing economy.

With intense competition driving suppliers toward thinner margins (less

173

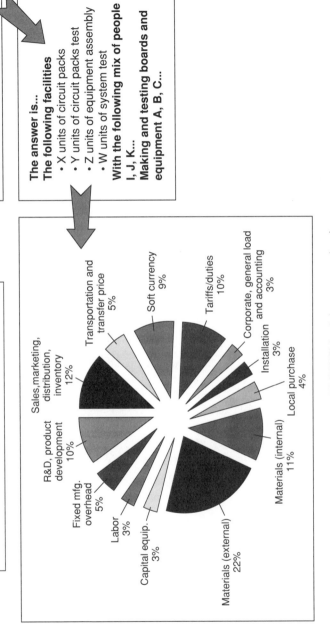

FIGURE 3 Joint venture planning.

margin or profitability), the inventory costs of finished goods in-transit also become a more significant factor in establishing global manufacturing locations. In fact, inventory carrying costs can offset some or all of the advantages obtained from lower labor rates. As a result, manufacturing companies will be driven to seek manufacturing locations closer to their ultimate customer even if it means higher labor costs.

When electronic manufacturing is established in developing countries, the supporting infrastructure and supply base will be motivated to move with it but at a slightly offset scale. The lack of a developed supplier base can significantly impact the competitiveness of a manufacturing operation when the lead times are lengthened and the inventories are increased to accommodate the need to import components not available in the local economy. The situation is further exacerbated when the physical and institutional infrastructures do not support consistent transportation intervals. In the global arena, competition occurs among full-stream production networks, sometimes called "value chains," and not simply between manufacturing locations. Consequently, developing countries need to understand the potential for the entire production network, not just the assembly operations in their country. Steps to foster a climate and infrastructure for suppliers—such as establishing industrial parks with the appropriate facilities, duty-free zones, simple import/duty regulations, and transportation—will make the transition to manufacturing more effective.

SUMMARY

In summary, rich information technology, globalization, and ever-increasing global competition are the key forces driving manufacturing technology trends. Recent improvements in manufacturing operations, supplier relationships, concurrent design tools and methodology, and the capabilities of the people who make it all happen should provide insight into the approaches to nurturing manufacturing in any country. If developing countries can foster an environment in which their manufacturing operations can realize the benefits of this changing landscape, they will be able to enjoy the economic effects of a viable manufacturing operation.

NOTES

1. Francis Stewart, "A Look at the Ups and Down," *Circuits Assembly* (September 1992): 25.
2. Arthur P. Cimento, Jurgen Luge, and Lothar Stein, "Excellence in Electronics," *McKinsey Quarterly* 3 (1993).
3. Ibid.
4. Jeffrey H. Dyer, "Dedicated Assets: Japan's Manufacturing Edge," *Harvard Business Review* (November-December 1994).

Innovations in Energy Technology

RICHARD E. BALZHISER
President and Chief Executive Officer,
Electric Power Research Institute

Over the next half century, the developing nations will vastly expand their use of commercial energy to meet the basic needs of their rapidly growing populations. But the drive for improved health, education, and welfare, as well as some measure of economic parity with the industrialized nations, will inevitably lead to pressure on the global environment, including climate. Not surprisingly, then, new, innovative means of global sustainable development will be required, with efficiency forming the backbone of all future strategies of sustainability.

Electricity will play a crucial role in fostering this innovation because of its unique ability to bring precision, control, and versatility to the workplace, and to capture and convey information. But while contributing to a less energy-intensive path, electrification also is among the most capital-intensive of the energy options. Thus it is imperative that electrification strategies crafted for each developing country consider its human, energy, and capital resources and focus on high-leverage applications that improve education and health, thereby enabling industrialization and economic development. Indigenous resources and practices will play important roles as nations seek to find their niches in the global economy.

This paper addresses the role that electricity can and will play in the rapidly globalizing economy. And although the paper concentrates on the new technologies available for generating and delivering electricity, the innovative ways that recently have been found to utilize electricity to improve economic productivity, reduce environmental degradation, and move toward sustainable use of global energy resources are where much of the excitement resides.

AN OVERVIEW

Energy use is driven primarily by population growth, economic development, and technological development. Worldwide, population growth is exploding—from 2 billion in 1900 to 5.5 billion in 1995 to perhaps 8-10 billion by the middle of the next century. Much of this growth will occur in the emerging nations where economic development is a top priority, and this development will require prodigious amounts of energy.

World energy use has soared during the last half century (Figure 1), with fossil fuels providing the bulk of the increase; nuclear and hydro (shown in the figure as primary electricity) have made minor contributions. The world now uses over 300 quads of energy a year, with the United States consuming a quarter of that total. Half of the world's people still have no real access to commercial energy, however. Efforts to improve their quality of life will be a major driver in future energy use.

Much of the growth in energy consumption over the next 50-60 years will be fueled by coal, oil, and gas, which are very unevenly distributed around the world. Coal is the overwhelming resource in Asia, Russia, and North America; the only significant supply of oil is in the Middle East; and gas is particularly

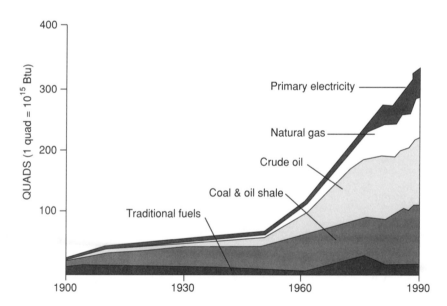

FIGURE 1 World energy use. SOURCE: Reprinted, with permission, from Donella H. Meadows, Dennis L. Meadows, and Jørgen Randers, *Beyond the Limits: Confronting Global Collapse, Envisioning a Sustainable Future* (Post Mills, Vt.: Chelsea Green Publishing, 1992), 67. © 1992 by Donella H. Meadows, Dennis L. Meadows, Jørgen Randers.

abundant in the Middle East and the Russian territories. Of the fossil fuels, only coal has staying power, although technology has gone a long way toward increasing the reserve-to-production ratios for both oil and gas in recent years. As for the solar resource, concerted efforts will be made to exploit this resource in the twenty-first century.

A look at the intensity of energy use reveals a number of historic patterns and phenomena. From the beginning of the industrial revolution to just after World War I, the United States went through a development cycle in which the energy input per unit of economic output climbed steadily. At the end of that period, however, the energy intensity of the economy turned down. Why? It was true that automobiles, furnaces, and industrial processes were becoming more efficient, but also electricity was taking root in the economy. In the 1920s, the electric motor unit drive revolutionized manufacturing. Later, it revolutionized the office, home, and farm. As electricity continued to replace less-efficient energy sources, the amount of energy needed to produce a unit of gross national product (GNP) steadily declined. Electricity's fraction of total energy is now approaching 40 percent in the United States and continues to climb.

This phenomena has not been unique to the United States; every industrialized country, beginning with Great Britain in the 1880s and extending through Japan in the 1950s (Figure 2), has passed through a similar historic pattern in

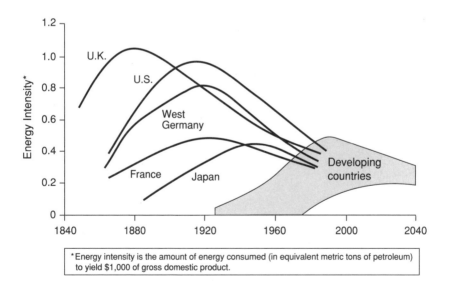

*Energy intensity is the amount of energy consumed (in equivalent metric tons of petroleum) to yield $1,000 of gross domestic product.

FIGURE 2 Energy intensities of industrialized countries. SOURCE: Reprinted, with permission, from Amulya K. H. Reddy and José Goldemberg, "Energy for the Developing World," *Scientific American* 263 (September 1990): 112. © 1990 by Scientific American Inc. All rights reserved.

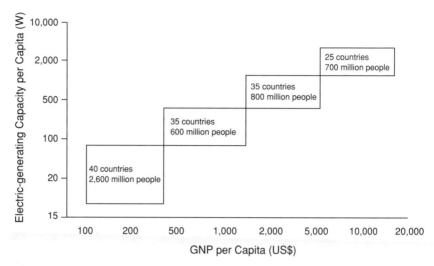

FIGURE 3   Prosperity and the use of electrical energy, 1985. SOURCE: ASEA Brown Boveri.

energy intensity in which, over time, the peak of energy intensity has declined. This pattern has enormous significance for the developing nations. In these nations, energy technology has advanced so rapidly in a historical context that they can pass through the same development cycle using only a fraction of the energy required a century ago.

Although the use of electricity has been closely linked to economic prosperity and improvements in social welfare, electrification has exaggerated the global disparities in energy use (Figure 3). More than a tenfold difference in electric-generating capacity per capita exists between the rich and poor nations, correlated with a nearly a hundredfold difference in per capita GNP. These gaps must be closed. But those working toward that goal must recognize that the type and scale of technology appropriate for the wealthiest nations may be very different than those appropriate for the poorest nations.

As the next century unfolds, the issue of global sustainability will begin to transcend the separate concerns of population, energy, economy, health, social welfare, and the environment. New means of achieving sustainable development will be required, and *efficiency* is likely to act as the backbone of all future strategies of sustainability. In this, electricity will be important in reconciling human aspirations with resource realities.

ENERGY TECHNOLOGIES

Of the trends important to the future direction of energy technology, efficiency is first and foremost. Fossil power plants were only about 5 percent effi-

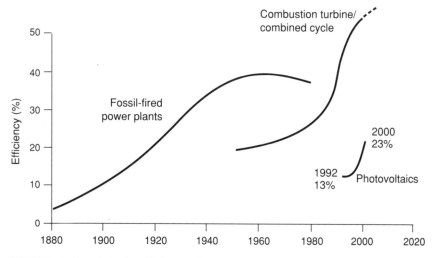

FIGURE 4   Trends in the efficiency of energy sources.

cient at the turn of the century (Figure 4), but today they are routinely 35 percent or more efficient, even with extensive environmental controls. Tomorrow they will be in the 60-70 percent range, with environmental cleanup integral to the process design. Similar improvements in efficiency are in store for the delivery of power as the "second silicon revolution" takes hold. And on the end-use side, one only has to look at the dramatic improvements in computers over the last few decades to see the potential for more gains in efficiency through electricity.

The wide variety of advance combustion turbines that are emerging will eventually push combustion turbine efficiencies into the 60 percent range. Aeroderivative combustion turbines, which, as the name implies, are derived from today's aircraft fan jets, are more compact than today's heavy frame turbines and are designed for rapid ramp-up to full power. They can achieve efficiencies of nearly 50 percent at the 50-megawatt power level. Advanced cycles of future interest for the aeroderivatives include the steam-injected gas turbine (STIG), the humid air turbine (HAT), and the chemically recuperated, intercooled, steam-injected turbine (CRISTIG).

The combustion turbine, which burns natural gas and is dominating new capacity additions in the United States, is readily transferable to any region of the world where gas or oil is available. The advantages of this turbine include high efficiency, low capital cost, low emissions, modularity, short lead time, and, at least for the foreseeable future, low fuel costs. These machines can be used in simple cycle or coupled with a heat-recovery steam turbine in a combined cycle mode. Because this technology has advanced rapidly, machines with capacities in excess of 250 megawatts are now available in sizes that permit factory fabrication and railway delivery.

Coal-based Technologies

Coal, which represents about 80 percent of the world's fossil reserves, is a resource so large and geographically dispersed that the world's dependency on it can only grow. In China and India, coal is destined to be the principal energy pathway for economic development over the next half century. Indeed, by one estimate, to accommodate its growth, China must build one medium-sized power plant every week between now and the end of the decade.

Fortunately, two decades of research on clean coal technologies stand ready to help minimize the environmental impact of coal. These technologies range from coal preparation to post-combustion control. But the real promise lies in technologies for which higher efficiency and environmental cleanup are integral to the engineering design.

Fluidized-bed technologies are particularly attractive near-term options for coal combustion in the 50-100 megawatt range. These systems trap sulfur in a limestone bed and burn cool enough to suppress the formation of nitrogen dioxide.

Perhaps the most significant technology for the long term is the integrated gasification combined cycle (IGCC), which gasifies coal, strips away the impurities in the gas, and then runs the gas through a combustion turbine in combined cycle mode. IGCC systems can be designed to convert coal into many different products, including electricity, chemical feedstocks, and liquid fuels. Advanced versions, now being tested on a pilot scale, replace the combustion turbine with a fuel cell, lifting systems efficiency to about 60 percent.

IGCC is beginning to catch on worldwide (Figure 5). Today, IGCC systems are being planned or under serious consideration in 21 countries, but this is just a start. Close observers of China's energy situation regard IGCC as China's best technology option for the long term. In the United States, the 100-megawatt cool-water facility in the Mojave Desert, first developed 10 years ago by the Electric Power Research Institute (EPRI) and a number of industrial and utility partners, has attracted visitors from many different countries, all eager to see the cleanest power plant in the world.

A comparison of the costs of the five currently available fossil systems that will set the competitive benchmark for electricity generation globally reveals that today's IGCC system can produce power at just over $0.04 per kilowatt-hour (Figure 6). (The kinds of cost figures shown in Figure 6 tend to be very site-specific, and the ones actually used represent some standardized assumption for the United States.) This is comparable to the figures for the traditional pulverized coal plants with flue gas scrubbers, but nearly 20 percent higher than those for gas-fired combustion turbines.

Fuel Cells

Fuel cells promise to be an especially clean and highly efficient form of dispersed fossil generation. Unlike conventional combustion-based technologies,

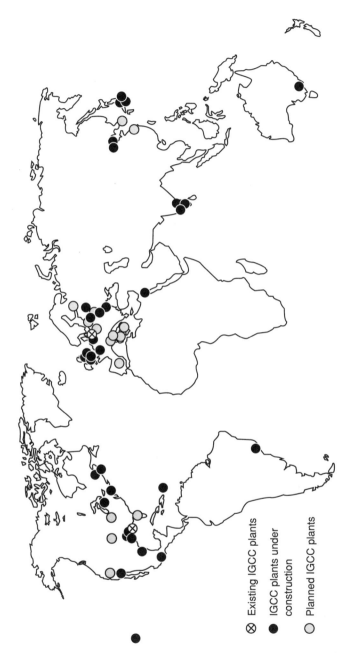

FIGURE 5 Global deployment of the integrated gasification combined cycle (IGCC) system.

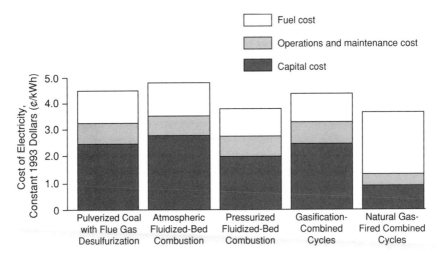

FIGURE 6   Cost of electricity from fossil fuel power technologies in the United States.

fuel cells convert fuel to electricity through a flameless oxidation process, much like a battery. Development activities are now focused on the molten carbonate fuel cell, with potential capital costs in the $1,500 per kilowatt range and a thermal efficiency of 54-60 percent. EPRI is participating in a 2-megawatt demonstration in Santa Clara, California.

Further out, researchers are excited about the potential of solid oxide fuel cells for extremely compact, low-cost dispersed generation. Inexpensive units as small as 5 kilowatts (or units as large as 5 megawatts) can be run at 50-55 percent efficiency. Larger cells can produce steam hot enough for industrial applications, boosting total efficiency to the 80 percent range.

Renewable Technologies

Photovoltaic solar technology should head the list of promising technologies for remote applications in the equatorial regions of the world. This technology is modular at scales appropriate for even low-power applications in village life. For example, the some 600,000 villages in India that receive 12 hours of sunlight daily could benefit from the use of "pre-electrification" levels of power for simple lighting and cooking, low-power TV sets, and, in some households, efficient refrigerators.

Even though photovoltaic systems cannot compete today for bulk power generation, they are often the least expensive option for small, distributed applications in remote areas, even in North America. The market is building as costs decline. Worldwide, photovoltaic sales are about 60 megawatts a year and growing between 15 and 20 percent annually.

The real long-term promise of photovoltaics rests on the fact that it is part of the family of solid-state technologies that are still in the early, robust stages of technological discovery and development. Research continues to make great progress: cell efficiencies are climbing; manufacturing techniques are being adapted from integrated circuitry; and costs are falling. In the last decade, costs have fallen from about $1.00 per kilowatt-hour to about $0.30 per kilowatt-hour, and within 20 years these costs are expected to fall even further to between $0.07 and $0.10 per kilowatt-hour.

Most of the focus has been on crystalline silicon flat plate technologies. But others bear watching, for they will extend the range and scale of future application. Thin films of amorphous silicon or other semiconductors have less demanding material requirements and hold out the promise of very low-cost, routine use. Their relatively low efficiencies can be boosted using multiple film "sandwiches" to absorb a broader spectrum of sunlight.

At the other end of the scale, high-concentration systems will become useful for utility applications. A 2-kilowatt installation now going up in Georgia is expected to produce power at 19 percent efficiency—a world record for total systems efficiency. This is expected be followed shortly by a 20-kilowatt system in Arizona able to produce power at about $0.10 per kilowatt-hour.

Solar thermal systems represent another intriguing option for utility-scale application. The most commercially advanced of the various designs is the trough-type collector. In the 1980s, California-based Luz International installed a number of these systems, using as backup a natural gas power generator. The systems were competing, however, in a size range that was the natural niche of combustion turbines, which burst on the scene as a fierce competitor in the late 1980s. Because the turbines could be installed for roughly one-tenth the capital costs of the Luz system ($3,000 per kilowatt-hour), the market failed to develop, and Luz went out of business a few years ago. But interest is rekindling in exporting solar thermal systems to regions of the world without gas supplies. To that end, the Rockefeller Foundation is now trying to pull together the various Luz subcontractors.

Wind energy is coming of age for power production. It has become competitive in the United States; it is rapidly taking hold in Europe; and now it is poised for broad international deployment. The developing nations, particularly India, have shown increasing interest as well.

Technologically, the leader in this field is the variable-speed turbine developed by U.S. Windpower, EPRI, and the U.S. Department of Energy. This turbine is capable of producing electricity for $0.05 per kilowatt-hour, given an average wind speed of 16 miles per hour. The turbines are rated at 350-450 kilowatts and can operate at wind speeds of between 9 and 60 miles per hour. Fierce competition is setting in, and a variety of new wind machines—constant speed and variable speed—should emerge in the next decade. Wind power costs will likely drop to the $0.03 per kilowatt-hour range within the next two decades.

TABLE 1   Cost of Electricity from Renewable Sources
(cents per kilowatt-hour), 1995-2010

Energy Source	1995	2000	2010
Photovoltaics	30-40	10-20	7-10
Wind	5	4	3.5
Bioelectricity	8[a]	7	5
Solar thermal	12[b]	10[b]	6-8

[a]Waste is delivered free at $0.04-$0.05 per kilowatt-hour.
[b]Gas-hybrid system (25 percent gas).

Biomass is a vast resource that includes waste products, as well as crops grown specifically for energy production. The fundamental challenge with biomass, however, is ensuring that its use to produce energy is compatible with other agricultural interests.

To produce energy from biomass, one can either burn the materials directly or convert them to a gaseous or liquid "biofuel" such as ethanol or methanol. Biotechnology is slowly reducing the costs of biofuels by opening up the cellulosic resource base. Waste from lumber and agricultural production currently constitutes the largest source of biomass, but many studies also are looking at short-rotation trees and perennial grasses that could be cultivated specifically for electricity or fuels. Moreover, new conversion technology holds out considerable promise. For example, the World Bank is sponsoring a project in Brazil to gasify a waste product—bagasse from sugar cane—to fuel an advanced combustion turbine. Gasification technology is important worldwide.

The costs of renewable energy are expected to decline substantially over the next 15 years and to reach a new plateau of competitiveness early in the twenty-first century (Table 1). It is conceivable that the economics of some of the more dynamic technologies, such as photovoltaic, could improve even faster. Certainly the incentives are growing, with the emerging markets for dispersed generation in Asia, Africa, and Latin America.

MODULARITY AND DISPERSED GENERATION

Renewables, gas turbines, fuel cells, and other future generating options offer the advantages of modularity—small, factory fabricated, and quickly installed. They can be placed close to the load, deferring the need to build and maintain transmission and distribution lines. As the costs fall, modular units are beginning to challenge the primacy of the central station design, and technical visionaries have begun to describe a time when the economies of scale will be superseded by the "economies of precision."

This will open the door to dispersed generation (Figure 7), which may evolve

185

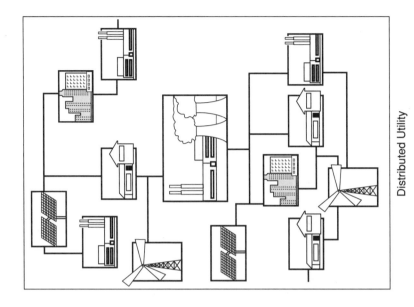

Distributed Utility

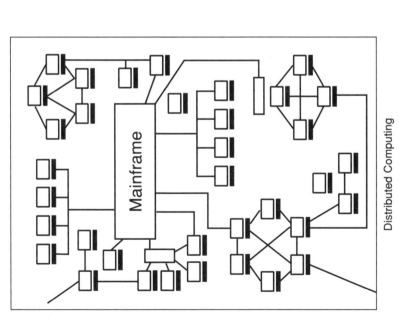

Distributed Computing

FIGURE 7 New paradigm: distributed production.

much in the same way the computer industry has seen large mainframe computers give way to small, geographically dispersed desktop and laptop computers. These can be stand-alone units, or interconnected into fully integrated, extremely flexible networks.

DELIVERY OF POWER

New technology also can assist in the challenges of transmitting and distributing power in the developing nations. In the poorer nations, the issues are primarily those of bringing electrification to the countryside—that is, of finding low-cost means of distributing power to often very remote and scattered villages and farms. This typically means running single-phase lines and contending with the myriad problems of maintenance and reliability. Some of the new equipment for condition-based monitoring, for example, could be very helpful in remotely sensing the conditions of breakers, transformers, and substations. With tools such as these, utilities can improve service and minimize the number of qualified technicians required.

At the other end of the spectrum—in the booming economies of some of the Pacific Rim countries—the transmission and distribution problems are somewhat different. Transmission lines tend to be loaded beyond their capacity from continuous growth in electrical demand, and serious power quality problems are emerging from new high-tech loads. The plasma arc of a new minimill, for example, can produce an instantaneous surge in demand of 300 megawatts, which is enough to break a generator shaft.

This is where the thyristor-based technology of the second silicon revolution will have a major impact. Macroelectronics brings the electronic switching capability typically found in microelectronics up to the half-million-volt level, effectively revolutionizing power delivery. Power can be sent to precisely where it is needed; network stability and power quality problems can be controlled; and long transmission lines can be strengthened. Power flow over lines longer than 100 miles is constrained by stability limits. An EPRI-developed Flexible AC Transmission System (FACTS) can provide the additional support required to improve grid stability. Indeed, by using FACTS, a utility can, with a relatively small investment, double the amount of power that goes down a given line. The investment in FACTS to double the amount of power in a 400-mile line, for example, might be as low as $20 million—representing as little as 5 percent of new construction. Given the phenomenal growth in the Asian markets, FACTS technology has a role to play in relieving the burden of new transmission line construction and improving electrical system reliability, which will allow the economies of these countries to flourish.

To appreciate electricity's unique role in the world and to envision the long-term future of energy technology, it is important to recognize that electricity has become the gateway to the electromagnetic spectrum. Gradually more and more parts of the spectrum have been put to work: radio waves for communications, X-rays for medical diagnostics, microwaves and infrared for heating. Now the world is witnessing an explosion of energy-efficient electrotechnologies for industrial use and public works: plasma-fired furnaces, radio-frequency and infrared textile drying, microwave drying of food products, microwave disinfection of medical waste, ultraviolet and electron beam treatment of wastewater, to name a few.

The high-tech world of today will become increasingly dependent on the ability to manipulate the spectrum. The direct result will be an enormous market pull for the efficiency and productivity gains available through high technology. And the indirect result will be better conservation of resources and a reduced environmental impact.

CONCLUSION

Electricity, whether 50 or 60 cycle or even direct current, can be converted to other forms or frequencies to provide high-value services from light to motive power, from communication networks to medical diagnostics and therapy, and from environmentally clean industrial processes to clean and efficient transportation systems. Just as the environment of many heavy industrial centers has been cleaned up using electric steelmaking and minimill technology, such urban centers as Los Angeles can be cleaned up using increasingly clean vehicles—internal combustion and electric—in this decade.

Heat pump technology continues to improve, providing increased efficiency and comfort with less use of primary energy. More specifically, heat pumps "concentrate" solar energy from the air or ground and deliver it to the home or office with overall efficiencies that exceed 100 percent in terms of energy delivered for each primary unit of fossil fuel used to generate the electricity.

Transportation, space conditioning (heating, cooling, ventilating), and cooking (where microwave and other new technologies are increasing their market penetration) are all potentially high-volume users of electricity in the developed world. For many developing countries, however, better lighting technology, water purification and waste detoxification electrotechnologies, single-phase motors, more efficient and forgiving electrical delivery and storage systems, and even increasingly efficient and affordable photovoltaic and wind machines for remote areas will begin the electrical revolution and will accelerate it for others wanting to speed along the path to a better future.

This being said, some observers have maintained that electricity is too elegant, too inefficient, and too environmentally burdensome to stay the course. These concerns deserve serious consideration if nations are to begin to demonstrate the global stewardship necessary to ensure that today's young people have enough resources to realize their aspirations for a better life.

First, to the point of elegance, electricity is an energy form and indeed the ultimate in elegance and versatility. It is quick (speed of light), clean, and efficient (90+ percent), and it can be produced from any of the natural energy resources—fossil, renewable, or nuclear—as well as waste.

The second point raises the question of efficiency and the historical reality that only 20-30 percent of primary energy can be converted into electricity, with the rest discharged to the atmosphere. Historically, fossil and nuclear energy have relied on the steam turbine to generate electricity. While this technology has improved significantly, the gas turbines currently available are able, in the combined cycle configuration, to convert gas, oil, and gasified coal at efficiencies of from 40 to 60 percent. These efficiencies will continue to rise. Renewable conversion efficiencies also are continuing to rise, and, as they do, the economics will improve because the fuel is either free—water, wind, and sun—or modestly priced—biomass or waste.

Finally, can the environment tolerate the emissions burden that 10 billion people will impose? "Innovation" is the most potent of the renewable resources, and it should be assumed that the future will hold a better understanding of and answers to the climatic concerns, nuclear waste disposal, and recycling opportunities that will extend the earth's finite natural resources.

In energy terms, it seems imperative that the viability of the nuclear resource be sustained and that the ability to use renewables and coal cleanly be improved. While electricity will likely increase its market share as the form of choice, direct heating, furnaces, gas stoves, and internal combustion vehicles and airplanes also are likely to be evident for a long time. Fuel cells will probably offer an opportunity to convert to electricity at the point of use whether in the home, the office, or the automobile. Gasified coal will back up the natural supply of oil and gas and will be used primarily for feedstock and electricity.

Educational Technology for Developing Countries

ALAN M. LESGOLD
*Learning Research and Development Center,
University of Pittsburgh*

For developing countries, as for all others, technology, especially information technology, is a tempting way to raise productivity. Just as new technologies have made it possible to produce better products at lower cost with less human labor, so it is often believed that technology can improve the learning process and perhaps even substitute for teachers who are not available or not affordable. Because they often have more educational chores to accomplish and less money to invest, the developing countries especially could benefit from the technological leverage of learning. But because educational systems are extremely stable and resistant to change, it is important to establish clearly whether a given technological contribution will sufficiently enhance educational productivity before undertaking any major effort to use it.

In this information age, the process of learning to use information tools may have value in its own right, beyond the value of those tools for promoting other learning. Technology can be a diversion, though, even a barrier to education, if it takes too much of a teacher's time to set up, or if it distracts teachers from productive tutorial interactions with students. This paper provides several different viewpoints on potential technologies for education and training in developing countries.

MATCHING EDUCATIONAL TECHNOLOGY AND CONTENT TO SPECIFIC ECONOMIC GOALS

Different situations and goals require different instructional and technological approaches. One key instructional requirement is to make optimal use of tools and curricula from outside while simultaneously respecting the need to build new knowledge on the existing local culture and knowledge. Educational efforts will fail unless they are attuned to the indigenous culture and its extant knowledge. For example, Patel[1] has shown that simple health-related measures, such as dietary changes to combat nutritional disorders and use of condoms to prevent AIDS and control fertility, are difficult to teach effectively without adapting to existing culturally-based knowledge.

The willingness of developing country governments to invest financially in educational technologies will necessarily depend on the scale of the educational goals being considered and their potential for economic return. It might be rational to spend thousands of dollars per trainee to produce a leadership corps for a high-yield industry that a country finds strategically central. But a developing country cannot afford high-tech solutions for universal basic literacy education unless that technology substantially improves the efficiency and yield of the educational process.

RECENT TRENDS IN ESTABLISHED TECHNOLOGY

In many respects, the problems surrounding the use of educational technology in developing country schools are similar to the problems surrounding its use in American schools. The U.S. educational system is remarkably stable and resistant to technology-driven change, just as the cultures of developing countries are stable and resistant to change. For example, schools, with their limited budgets and complex procurement systems, find it difficult to handle the materials, labor, and consulting costs related to computer technology maintenance, and teachers have limited time to invest in learning new instructional methods. Furthermore, in each case the extra dedication required of teachers to surmount logistic difficulties will be forthcoming only if the positive effects of the technology are clear and substantial. This leads to three basic principles to follow in adopting education tools.

The first principle is that *low-maintenance commodities work best.*[2] The American school scene is replete with horror stories of computers sitting idle in closets because the batteries for their clock chips (and parameter memory) died or because there was no money to repair a hard disk that was dropped. Developing countries face similar problems. Thus the best technologies will be those able to function with little or no maintenance (this is not an absolute rule but rather a guideline). Although a countrywide center for training software designers[3] or an urban training center for health paraprofessionals might profit from advanced computer technology, a village removed from urban centers might do best with "disposable" wireless phones or radios combined with desktop-published text materials shipped from a regional production point.

Word processors, spreadsheets, symbolic mathematical manipulation tools, and information network clients are among the software tools that are available as commodities. American school systems often decide to purchase special-purpose versions of tools even though generic products with similar or greater capability are available widely in commodity form—the commodity version may have appeared too complex, or the special version may have offered some feature that seemed useful. Commodities, however, are cheaper by orders of magnitude and much easier to maintain. Moreover, an economically powerful customer base supports the maintenance of commodity software, by both purchasing technical support and exercising purchase choices, whereas schools rarely have enough money to pay for good technical support. Some universities, however, have been quite successful in negotiating extremely low-cost package software deals for their students, such as complete office packages (word processor, spreadsheet, database system, and so forth) for prices below $100.[4]

The second principle is that *easy-to-use tools work best*. When a piece of technology is difficult to use, and especially when it is difficult to *start* using, it has a much lower chance of penetrating a culture to which it is foreign. The slow-changing cultural lore needs time to recognize that it will be worthwhile to master a new technology.

Even simple technologies can be hard to use when their usage is poorly staged. For example, often in U.S. classrooms a teacher wishing to show a video must first locate the massive, unwieldy cart containing the video display monitor and then roll the cart through crowded hallways to his or her classroom. No wonder video is not used as much as it could be! The specific barriers to easy deployment of technology may differ in developing countries, but the same principle—that difficult-to-use devices will not get used—certainly applies universally. Moreover, the threshold for ease of use is quite high. Devices that are not readily accessible to all who approach them will not work for mass education.

The third principle is that *tools that foster activities that promote learning work best*. This rule may seem obvious, but it is not. Too often tools that seem "educational" do not lead to the learning of anything important. Indeed, a standard marketing strategy in the United States is to claim that a particular piece of software, such as a computer game, is "educational"—perhaps it teaches higher-order thinking skills. While it is productive to use such commodity tools as spreadsheets, word processors, and information server clients (for example, Mosaic) to support classroom activities, it is generally not as effective to invest student and teacher time in learning to use a program when that program supports learning only incidentally. Many computer games fall into this category, as do other systems that take a long time to learn to use relative to the amount of learning they can support.

TWO EXAMPLES OF EDUCATIONAL TECHNOLOGY

How then do these principles apply to specific technologies? Two interesting examples of technologies are described here. Interactive radio, a low-cost scheme that has been used increasingly since 1974, is included as an example of educational technology deployment because of its substantial success, at least for the limited purposes for which it has been used. The second example, which is at the other extreme of technology, is intelligent computer conversations with students. Using new and powerful intelligent systems technology, this option has considerable potential as an approach that is affordable and distributable in developing countries.

Interactive Radio Instruction

Interactive radio is an idea that most educational technologists would not favor because it is not really interactive.[5] Lessons are spoken over the radio, and students respond to prompts from the speaker; nothing the student says is preserved or transmitted to the speaker. Why would any educational technologist want to develop such a system? First, data show that interactive radio is remarkably cost-effective for schooling. Second, radios are ubiquitous and exist in a highly sustainable form. They also are cheap and are easy to maintain; require only batteries or central electrical power; and are easy to operate. A basic radio has only three control functions, usually lodged in two knobs: system activation (on/off), frequency tuning, and volume level. If standard broadcast bands are used, tuning is extremely straightforward. And, most important, many world cultures support the use of a radio. Thus students, teachers, and other learning mediators (for example, parents) already know how to use them.

In most cultures, a considerable amount of learning activity involves verbal interactions. Such verbal interactions are costly, however, since they depend on a teacher, who can deal only with a small number of students at a time. Because most developing countries have too few qualified teachers, they especially would benefit from technological approaches that replace at least part of what a teacher does. For example, it is simple to replace the set of learning conversations in which a teacher initiates each section of a conversation and in which the conversation can continue when the teacher keeps uttering statements that are not determined by what the student has said. Teachers can provide examples, espouse principles, and even pace the progress of students through linear problem solving without necessarily having to react to the specifics of what students say in reply. For example, a teacher can ask a question, pause for an answer, and then explain why a particular answer is good. In addition, when a group of students listens to radio-delivered "interactive" instruction, each student also learns from the responses he hears his friends calling out alongside him.[6]

Tilson has reported on a number of studies of the effectiveness of interactive radio in elementary education.[7] One study conducted in Bolivia around 1989 compared students using interactive radio as part of instruction in mathematics to others using only standard textbook instruction. Some of the former also had used interactive radio the previous year for second-grade mathematics. The differences in post-test performance between the two groups were substantial (Figure 1): effect sizes of 1.29 and 2.03 standard deviations for one and two years of radio-delivered instruction, respectively.

In a range of similar studies (but for the second grade) conducted in Bolivia, Nicaragua, and two areas of Thailand, effect sizes of 0.24 to 0.94 standard deviations were obtained for interactive radio. Other ages and other topics also have been investigated. Health education studies have shown that the radio approach is particularly effective for teaching upper primary students the details of oral rehydration therapy delivery and other aspects of gastrointestinal infection in infants.

The interactive radio approach is relatively inexpensive to start up and to sustain. Costs estimates range from a high of $6.40 (the cost per student in 1990 for serving 25,000 students in Bolivia) to a low of about $1 per student to sustain larger efforts once initial capitalization has been completed. These costs are not much out of line with the most cost-effective approach to basic primary education—textbooks.

Interactive radio is not effective for all instructional needs. For more difficult instructional and training needs, interactive radio is not interactive enough or individualized enough. It works exactly where information and prompting can be provided unambiguously and with minimal regard for moment-by-moment changes in students' understanding. Its social effects, however, should not be

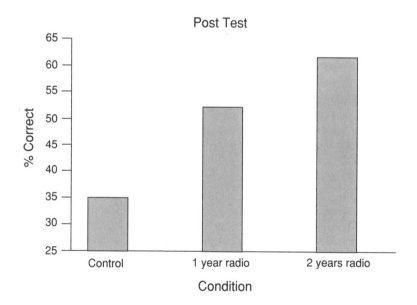

FIGURE 1   Interactive radio effectiveness (post-test) in Bolivia, third-grade mathematics.

dismissed (students are part of a learning group that listens to radio together), and it provides effective delivery of factual knowledge, as well as a social comparison base for students who can watch each other's responses.

Because radio also depends on having little or no need for pictorial or diagrammatic information, it would not be the best way to teach physics students how to draw force diagrams or to teach technicians how to operate an intravenous pump in the critical care unit of a hospital. Thus the advantages of interactive radio are very limited for all the reasons that support the use of multiple media in education. Even so, the idea of distant, audio instruction could be refined considerably in the light of recent progress in speech understanding by computers.

Intelligent Coaching of Oral Reading

A few years ago, Jack Mostow[8] at Carnegie Mellon University proposed to build a computer-based coach for elementary school oral reading. Students learn to read from a combination of being read to, being coached on specific symbol-sound correspondences, and especially reading aloud to a teacher or parent. Mostow reasoned that even though the general problem of how to configure a computer to understand speech is not wholly solved, the problem of understanding a child who is reading a known text is far less complex. Basically, it involves verifying that consecutive words of a fixed script are in fact uttered, which is much easier than deciding which word a person has uttered when there are no constraints on the possibilities.

At present, Mostow has completed demonstration versions of an oral reading coach and is building a production version. The system can recognize mistakes and can prompt students when they get stuck. This tool, which does a lot of what a good teacher would do, is needed both in the United States and in developing countries. Equally important, the basic idea that speech recognition is tractable whenever the set of possible utterances is highly constrained can be taken much further. Students can be offered choices to which they can respond orally, thus allowing a real interaction to take place in the audio realm.

The approach is limited to cases in which there is a describable set of alternatives from which the student can choose. The simplest possibility is multiple-choice responses. While this is not the instructional form of greatest interest to educators, it has its uses. Simple choices that can be offered include self-assessments by the student. For example: "Do you understand why you must boil the water you use for oral rehydration therapy?" "Can you see why it is important to give a child with bad diarrhea liquids right away, even if he has a problem that your friends usually treat with dry foods?"[9] More complex choices might be possible if the student is able to handle occasional follow-up questions from the computer (students can handle partially deaf teachers, so they probably will adapt to computers that are not perfect in what they "hear").

Multiple computer-generated voices, different sounding voices for different roles, are another possibility. This would allow coaching or other feedback to come from multiple sources. For example, a training program in public health could include coached feedback from a scientist coach, a nurse coach, and a politician coach. Thus in an exercise dealing with an epidemic, a proposed quarantine might produce praise from the nurse coach but expressions of horror from the political coach.

While the value of intelligent conversation as a source of learning is clear, it will not be easy to move powerful computers to rural sites in developing countries, nor is it likely that local villagers could maintain these machines. The solution: the student and the computer, which would be located elsewhere, can talk by telephone (see Note 9). Soon it also will be possible to use relatively low-cost wireless communicators that have low-bandwidth video capability.

What determines the potential suitability of a technology for instructional use in a developing country is not the technology considered *in vacuo*, but rather the technology properly adapted to this purpose. Many technologies that appear too exotic for the developing world can, in fact, be adapted if the criteria of ease of use, cost-benefit ratio, and sustainability are considered carefully in the design of training or learning systems. And, equally true, many technologies that seem ideal can work out poorly if these considerations are not given careful attention.

WHAT MAKES LEARNING HAPPEN?

In the past, learning was seen primarily as data transmission, with perhaps a second component of practice. Recently, however, instructional researchers have become more aware of the constructive character of some learning. Especially when the material being taught is not part of the student's cultural background, learning must be a more active process—one in which the student constructs new knowledge in terms of prior experience. Effective tools for this kind of knowledge construction process either facilitate the student's efforts after understanding or force the student to confront gaps in his existing knowledge.

Each of the ways to foster learning (an incomplete list is shown in Table 1) has its own strengths and weaknesses. For example, learning by being told works only if the teacher and student agree on the meaning of the language that the teacher is using. Even then, this approach encourages confusion between what was said, the desired learning outcomes, and what the student knew earlier and learned later. No one learning approach is best for all circumstances. Thus it is best to develop a clear set of learning goals before selecting a training approach.

Of the two basic types of learning model described in the rest of this section, one type captures the highly circumscribed approach needed when the goal is to form a highly reliable (but perhaps somewhat redundant) cadre of workers able to handle situations routine enough that modest natural selection among workers is sufficient as an approach to the need for qualified practitioners. The second model captures at least the veneer of the "smart worker" concept.

TABLE 1 Forms of Learning

Mode of Learning	Comments
Being told	Requires prior agreement on the meaning of terms and a lot of relevant shared experience. Encourages confusion between covering material and successfully teaching. Obfuscates some of the responsibilities of teacher and learner.
Studying modeled expert activity	Excellent preparation for doing the same thing the expert is modeling. Good goal setting by example. No guarantee of transfer beyond the modeled activity in the situation in which modeling occurred.
Discovery	Tends to anchor acquired knowledge in relevant experience. It can be a long wait before some students will make certain discoveries. Not necessarily suited to assuring transfer.
Legitimate peripheral participation	Ideal for easing a person into a community of practice. Good motivation of learning activities. Affordances for novice "doing" may not match learning requirements.
Coached apprenticeship	Performance demands all occur in relevant situations. Transfer opportunities can be handled via simulation. Need to add something to deal with abstraction requirements for learning. Highly motivating and "honestly" motivated.
Reflection	Ideal for abstraction learning, especially if combined with approaches that provide extensive experience in real performance environments. By itself can lead to too much abstraction away from real situational requirements.
Collaboration	Assures a greater realism of acquired knowledge—negotiated meaning. Assures greater ability to talk about work and thus to work in a team. Both natural task-derived motivation and additional social motivation.

The Beehive Model

Kevin Kelly,[10] citing the work in robotics by Rodney Brooks,[11] has discussed the factors that make biological systems robust. These include parallel, redundant, and highly accurate subsystem modules. These modules are not completely programmed and are somewhat idiosyncratic, with the most adaptive being selected for a given situation. For example, a beehive feeds itself through this combination of reliability, variability, and natural selection. Bees forage somewhat randomly, but whenever a bee finds food, it returns reliably to the hive and does a highly programmed dance to which the other bees respond in a highly programmed way. In this manner, the hive optimizes its search for food without having to understand how it finds the food.

How would one build work teams that have this property? First, a subset of the skills needed for work would be highly drilled and practiced. Second, when novel situations arise, different team members would try different schemes to deal with the novelty, and a highly disciplined decision process would determine which scheme is to be supported. This approach has been partially adapted by some traditional workplaces to handle, for example, vigilance and safety requirements. If any member of a team notices a safety hazard, the entire team reverts to drilled procedures to preserve life and property. Every organization has some beehive-like functions. As a general approach to work force competence, however, the beehive model has serious shortcomings. Without a deep understanding of work situations, teams may not come anywhere close to having this redundant, reliable, but limited capability.

For those situations in which the beehive approach is feasible, certain forms of training are appropriate. A limited set of skills must be highly practiced, and motivational components must be included in the training regimen because of this overlearning requirement. Traditional computer-based drill may be sufficient in a few cases, if it can be tied closely to the situations in which the skills will be used. Richer, more visual, and more concrete simulation environments may be needed. Such approaches as spoken computer responses to trainee actions may be more motivating, and it is likely to be easier for trainees to work in teams, either at a single learning station or over a network.

Even in the area of traditional drill and practice, then, technology might augment performance by providing more realistic practice situations and by affording more motivating real and simulated interpersonal interactions.

The Smart Worker Model

Training and learning must be broadened for the work environment of the future. While the specific requirements for developing countries will be partly a function of the form of each country's participation in the world economy, it is possible that a smaller country will be best served by aggressively pursuing a few niche markets. Competitive advantage will stem from a combination of standard quality-enhancing practices and deep expertise in the niche domains. Each worker will need to understand the goals and functional characteristics of his or her enterprise and will need to adapt continually to different situations. To prepare a cadre of workers for high-productivity penetration of a niche market, one must offer a combination of skill development, practice with complexity, and adaptive problem-solving capabilities.

Based on my own training development experience, the combination of coached apprenticeship and post-performance reflection is especially productive. Apprenticeships allow basic skills to be exercised in realistically complex situations, while reflection on one's performance is the key to generalization and adaptation. Technology can play a role in both apprenticeship and reflection. For example, a system we built called Sherlock provides a simulation of a work environment (specifically, diagnosis of faults in a complex electronic switching system), an intelligent coach that provides advice when the trainee reaches an impasse, and a collection of reflection tools that allows trainees to review their own performances after solving a problem and compare their performances with that of an expert. Trainees also are given access to descriptions and explanations of system function and diagnostic strategy. Like other aspects of computational technology, the cost of building such training systems is dropping rapidly as the basic techniques are perfected, and especially as object-oriented design tools and object databases become more commonplace.[12]

The combination of apprenticeship using real or simulated work places with reflection on performance is essential if workers are to be able to deal with emergent situations. Developing countries, like those further along, need smart workers who can reason beyond what they explicitly have been taught. The "general" capabilities that can be taught, however, are not purely abstract; rather, they are grounded in one's experience and in real job situations that span the likely range of situations encountered in real work. It is difficult, perhaps impossible, to generalize from specific experiences while immersed in them. Yet it is not likely that abstract "book learning" will be available in usable form when concrete work situations demand that a worker stretch his knowledge beyond that which has been fully and explicitly mastered. Reflection, comparison of one's work to that of others, and discussion with colleagues are key means of acquiring stretchable knowledge. Technology for learning that supports these approaches will be particularly useful.

SPECIAL NEEDS FOR WORKER AND INSTRUCTOR TRAINING IN DEVELOPING COUNTRIES

Developing countries, like all countries, need to provide their citizens with a high-quality education. But this is a major economic burden to every leading economic power, and it is especially daunting for newer economies.

In addition to covering the traditional subject matter, schools should coordinate their academic goals for students with the skills and education needed by a modern labor force. Even though some advanced countries are studying the cognitive skill requirements for modern productive work, there remains a cultural legacy, too often copied in developing countries, of detaching schooling from work. While rote training for today's jobs, which will not last a lifetime, is foolish, so is abstraction of schooling to the point where it has no ties to past or future life experience. Alfred North Whitehead noted long ago that knowledge acquired in the sterile academy is often "inert," not coming to mind when relevant to real world situations. This never was a good thing, but it is particularly problematic in an era of rapid change in the nature of work. Because of the rate of change, the culturally shared knowledge about the workplace that has compensated for the removal of school from work life is decreasing. At the same time, a large cadre of people—teachers—have been successful in life by doing just what school people told them to do. It is essential, then, that teachers better understand the workplace and the ways in which schooling can facilitate learning to do specific work. Technology can help with this by bringing information about and simulations of modern work into the school.

Technical Training before Entering the Work Force

Today, productive work usually requires job-specific training beyond general schooling. In some countries, this takes the form of formal apprenticeship programs, usually combining on-the-job practice with classroom time. In others, apprenticeships are less formal or even nonexistent, but prospective workers still must have specific training before they are likely to find jobs. Too often, programs focusing on specific job performance have provided very specific practice, often on outmoded equipment, combined with totally abstracted schooling. For example, a prospective machinist might receive courses in trigonometry and mechanical drawing combined with exercises using a manual lathe, turning relatively cheap and workable metals.

The problem with such schemes is that in periods of rapid change, intuitions of instructors about real work situations do not change fast enough for them to revise the courses and practice opportunities to support the new jobs. A modern machinist, for example, works by programming a computer-controlled machining cell, or perhaps she receives the program from a programmer and checks to see if it is feasible in its sequencing of cuts and consistent with the target product. Furthermore, depending on the value of the metal and the cost of labor, critical concerns about the reliability and duration of the planned sequence of cuts will vary. Some machinists produce dozens of identical items from cheap metal, while others may spend days planning and executing the milling of a single piece of stock costing $250,000. Do math courses and practice making candlesticks prepare one adequately for such work? While no program is quite that antiquated, few current programs reflect sufficient thinking about the connection between training and subsequent work.

It is costly, however, to completely retrain a worker every time a new technology appears. Conceptual support for broader job families would certainly be worthwhile if it could be achieved. Jobs keep changing, it is argued, and thus students should receive broader preparation, but few programs are based on any systematic analysis of the range of likely jobs or of the key abstractions that can facilitate transfer from one job to the next. Recently, I suggested that the same ideas used to develop abstractions of computer software modules and to organize such modules for reuse also be applied to the systematic specification of knowledge requirements for training.[13] Specifically, just as separable pieces of a computer program, called objects, can be organized into abstraction hierarchies (usually called *inheritance hierarchies),* so knowledge objects can be specified and similarly organized. In such inheritance hierarchies, a more specific version of an object *inherits* many of its aspects from a more general "parent" object but also contains the unique knowledge needed to adapt to a special situation for which it has been developed.

Such hierarchies of knowledge specifications provide both computer and human tutors with a formal basis for coaching that is focused both on situation-specific competence and on broader understanding. The logic of this kind of coaching is to (1) use past experiences to establish a basis for generalization, (2) then introduce the generalization, and (3) specify what is different about the current situation that requires a specialization of the general approach.

Training can be made much more efficient and effective by using the object-oriented analysis and design approaches that have enhanced software productivity. After all, building a base of reusable knowledge objects to support a domain of software application is really no different than building a clear specification of the job-specific and more domain-general knowledge requirements of workers who are to be trained.

Training for New Tasks

A learning-by-doing approach also has many advantages for job-specific training.[14] This approach ensures that the knowledge needed is acquired in the context for which it will be needed. Through the use of computer-based simulations, it becomes possible to provide learning-by-doing training even in many areas where it might be impractical or dangerous to allow workers to do the real job before the proper training. Through the use of intelligent coaching, trainees can engage in rich work with all its complexity since the computer coach ensures that the trainee will not reach an insurmountable impasse; rather it converts impasses into learning opportunities. Through the use of post-task reflection, during which the trainee receives critiques of his performance and learns more about how an expert would have attacked the task, computer-based training systems also lay a groundwork for the trainee's subsequent transfer to new jobs.

Improvements in object-oriented analysis and design technology are key to building practical learning-by-doing systems, as well as the availability of relatively powerful desktop computers.[15] An infrastructure for such schooling can be adapted from prior efforts to introduce learning-by-doing and information management tools for training a vanguard work force for two or three niche markets. India, for example, will soon be able to build on the base of one of its niches—an applications software industry.

A Technologically Literate and Innovation-Receptive Population

Anyone seeking to exercise democratic citizen rights in complex societal decisions, to understand economic policy, and to participate in high-quality manufacturing, service, or information work, will require some new knowledge. Whether considering a public issue, such as to ban or not to ban the use of certain hydrofluorocarbons, or doing one's part in maintaining efficient, quality manufacture of consumer goods, people need increasingly to have tools for understanding, examining, and criticizing complex systems. Although developing countries can begin to improve their economic standing by training a small elite work force for a few target niche industries, further growth also will require a populace that understands the modern world and is prepared for modern productive work. But both the complexity and the interconnectivity of systems will pose learning requirements for schools. Students need extended practice in understanding and manipulating systems, talking about them to one another, and envisioning their function from multiple viewpoints, as well as the tools for managing information complexity. Partly because they may lack a universal analytic verbal tradition, the cultures of many developing countries may be better able to assimilate broad systems-based understanding of the natural and technical world than the cultures of the more developed countries. Still, tools for thinking—the kinds of outline tools, electronic lab notebooks, and simulation packages beginning to penetrate schooling in the richer countries—will eventually be important for developing countries. This will require not sweeping calls for two computers in every hut but rather educational goals aimed at competence with complex systems as the needed tools become maintainable, cheap, and easy to use.

REQUIREMENTS FOR SCHOOLING IN DEVELOPING COUNTRIES

Adapting to Local Language and Conceptual Structure

New systems of knowledge are not acquired independently; rather, they are layered on top of one's prior knowledge. For example, Patel found that even after years of immersion in U.S. cultures, women from South India still maintained a knowledge of nutrition that began from the folk knowledge of the regions where they were born.[16] Without understanding the systems of knowledge those women started with, researchers found it difficult, perhaps impossible, to determine what they understood and knew after exposure to European-American nutritional ideas. Similar situations arise when people from North America go to western Asia. For most Americans, the gastric acidity triggered by spicy food is an illness, to be fought with antacids if it becomes too noticeable. Such an attitude overlooks the role played by food in triggering bodily protections against bacteria.

This then is another requirement for efficient and effective instructional design: the explicit building of connections between what people in developing countries already know and certain bodies of knowledge they need to acquire to be productive. The new knowledge must be anchored in what is already understood. This will be the case especially when issues of health are involved, but it will arise in many other situations as well. In commerce, for example, different cultures view the relationships among workers, owners, customers, and managers differently. In more extreme cases, cultural differences in time perception, for example, become important.

Concreteness

One partial antidote to the cross-cultural communications problem is concreteness. Because verbal telling fails when the terms in the message are not understood, demonstration in a real or simulated world often can overcome such misunderstanding. A major consequence of the exponentially expanding bandwidth of computation and communication systems is the ability to replace symbols with enactments and to replace words with pictures, movies, and manipulable displays.

Developing a Teacher/Instructor Training Infrastructure

The developed countries may have a shortage of teachers who understand the modern workplace, but in developing countries teachers who have such an understanding may be nonexistent. Such a situation calls for development of a cadre of vanguard master teachers with a better understanding of the connections between what they teach and the kinds of jobs their students will have in the future. Here again there is potential for using workplace simulations that are provided and coached by a computer system, this time for use by both teachers and students.

A grass-roots, self-help capability among teachers could be fostered by improved communications capabilities. At the lowest level, telephone interactions can be useful in bootstrapping an improved teacher corps. Modest electronic bulletin board and electronic mail capabilities also can be very powerful, although there will be cultural differences in the extent to which a teacher feels it appropriate to pass on his or her ideas to others or to ask others for assistance. Modest levels of access to information servers containing workplace simulations would be a valuable addition and will become feasible as network technology becomes ubiquitous.

Assessment

Schools needs benchmark quality standards for the education process and for ways of measuring progress in attaining those standards. For developing countries, the only educational outcomes that really matter are those that move people significantly closer to being more productive. This requires continual assessment against clear standards.

Some aspects of the American approach to testing are not likely to help developing countries; for example, U.S. assessment schemes are grounded in a fundamental distrust of the student. In most total quality systems, measurement and benchmarking are possible largely because employees measure products themselves. If a company with a strong total quality management (TQM) program were to announce that it was hiring a major testing company to do all benchmarking rather than trusting its own employees, the effort would likely be a disaster. Total quality management works in large part because employees are trusted to make good assessments. Indeed, many aspects of quality control require employee input. Surely this must be true for quality control in schools as well.

In most U.S. schools, testing is adversarial, with an implicit assumption that the student's self-assessment is not to be trusted. This is unfortunate because often the student is in the best position to judge her own understanding and competence. To the extent that instruction is delivered by computer systems, those systems can be simpler and more effective if student self-assessment is part of the design. Students will need help in assessing their progress, however, because improvements from day to day in such skills as writing, painting, and other communications competencies are not obvious. (My own experience with trying to learn oriental painting supports this conclusion. I spent several years at it, with no sense at all of any progress. Only when I discovered a collection of old paintings my wife had saved over the years could I see that I had been improving slowly over time.) A major role of assessment in schooling is exactly what it is for industry, to help the student notice changes in the rate and quality of production. But that is only possible if the educational system and the student trust each other.

More broadly, students need to appreciate that their many investments in learning over a period of years, even decades, are headed toward valued goals. Indeed, students need to be able to see the consequences of their learning progress over the long term.[17] How else can a student in fifth grade decide whether she is working hard enough to be on track for an eventual job as an engineer? Social comparison and culturally embedded work norms are fine for highly homogeneous cultures in times when jobs and the training they require are stable. Too often today, however, schools serve a mix of students from multiple backgrounds—black versus white, urban versus rural, Ibo versus Yoruba—who do not trust each other's norms. Consequently, self-assessment tools that help the student to chart progress toward goals he values are needed.

Focusing on Areas of Greatest Need

Developing countries cannot afford to follow the lead of the United States on such issues as the extent of university resources to be put in place. Even in the United States, far too many students are receiving a low-quality "liberal" education that does not prepare them for work later, but this is being remedied somewhat by a significantly increased investment in postsecondary technical education.

Developing countries face hard choices in deciding which parts of an educational system to build first. It seems sensible, though, to develop a technical education and worker force preparation capability first, deferring the development of elite professional institutions until later. While the latter are important to the maintenance of a country's identity, they are also the most expensive to build, the hardest to maintain at high quality, and the easiest for which to substitute billets at foreign universities.

EDUCATIONAL TECHNOLOGIES WITH POTENTIAL FOR DEVELOPING COUNTRIES

Some technologies that offer a mixture of new ways to convey information and to support thinking have particular potential for education in developing countries. Furthermore, these technologies soon will be available as affordable commodities.

Combinations of Desktop Publishing and Printing Technologies

The tools for desktop publishing are improving rapidly. Word processing systems are increasingly able to produce hypertexts and hypermedia "shows," and virtually any enterprise can carry out the entire publication process using a combination of powerful word processing software, desktop publishing and presentation software, and new hardware/software combinations for printing. In a high-technology economy, print would be largely eliminated, replaced by network connections and access to such software as Mosaic.[18] Combinations of printed materials and minimal additional technology will continue, however, to be a primary means of conveying information to rural areas of developing countries. New, low-cost, computer-based printing systems, such as that produced by

Riso in Japan, offer a direct interface between standard Microsoft Windows print drivers and the production of printing masters. Thus a user can give the same print command that might lead to laser printer output and instead receive a master for a low-cost, ink-based printing system. The total system is easy to use, maintainable, and as economical as a commodity business desktop computer. On the printing side, for intermediate quantities (a few thousand) costs are much lower than for xerographic reproduction. The information revolution is improving the economics of traditional paper media production.

CD-ROM-Based Multimedia

Commodity distribution vehicles that can convey voice, motion picture, animation, or other information are a significant breakthrough. One can argue over how long the current CD-ROM will remain standard, but the costs for this technology are now low enough that many exciting educational possibilities exist. Moreover, compared to magnetic recordings, CD-ROM disks are more robust and can survive a much wider range of environmental exposure.

One particularly important side effect of the CD-ROM revolution and related technologies is the rise of a low-cost video production capability. Whereas video editing and production equipment was once so expensive that it rented for $100-$300 an hour, one can now build systems sufficient to produce decent video and deliver it on CD-ROM technology for $20,000 or less. With CD-ROM drives selling for as low as $250, the range of situations for which video is affordable is growing rapidly. Even if only a few video segments are needed to make an instructional system more effective, a central production facility for a country can afford to produce such products.

While the production of single CD-ROM products is quite inexpensive, mass production is somewhat more expensive. A likely scenario is that a developing country would buy some of its training software from elsewhere—perhaps from a country with a niche market in educational media production—and then adapt the software for local use, fitting it to the local culture and the experience of students in that country. The country's central educational production facility would then produce a single CD master that would be sent to an industrial producer for mass production.

Distributed Information Network Schemes

The worldwide information server network that has developed over an extremely short period allows students almost anywhere in the world to access information from almost anywhere else. Although it is not yet clear how long free access will continue, given a bit of luck straight access for educational purposes may continue to be low in cost or free, with entrepreneurial efforts focused on the problem of searching for and organizing information rather than controlling information pipelines. Even if much of the Internet becomes costly, however, it will remain feasible for developing countries to build local capabilities and to share educational content text, images, video, sound, and so forth via network commodity tools such as those initially distributed by the University of Illinois.

A substantial range of instructional possibilities is available using Mosaic because it supports a variety of still and moving image forms, sound, and a limited interactive capability, using text forms. Further flexibility arises from its ability to invoke a program as the response to a request for information—that is, a form submission or a hypertext link can point to a program on an information server that will be executed before any information is returned to the person making the request. This means, for example, that the World Wide Web capability is readily enhanced to include intelligent interactions between the information server and a student requesting information. In my own work building intelligent coached apprenticeship systems, I have found that most of what my system does could be delivered over the Internet since the student response in my system, while it always seems to be rich enough to mimic real work, is actually limited to selections made by pointing to menus and images. Indeed, we are currently pursuing several projects to convert tutorial systems to a client-server model that is based on the World Wide Web of information servers.

A variety of instructional technologies have been developed to take advantage of the basic network delivery schemes and hypermedia representation schemes that have appeared. At present, researchers in companies and universities are designing "shells" for the authoring of "learning-by-doing" systems, and another cadre of technologists is working on improved distributed information network technology. Simple, locally tailorable technology for distributed use and support of learning-by-doing technology will likely be available within the next few years. Combined with the rapid growth of wireless telephony technology, this will enable the establishment of powerful educational networks in developing countries. The vanguard of educational networks will likely be the laptop computer with a CD-ROM drive and wireless modem, but combinations of satellite and fiber-optic transmission probably will soon follow.

Affordable Basic Artificial Intelligence Tools

Rule-based programming already is playing a role in training design. For example, a system developed by National Aeronautics and Space Administration (NASA)—CLIPS—helps the user to conduct a job analysis. The analysis can then be formalized as a series of "productions," or conditional actions. This set of IF-THEN rules constitutes a simple form of "cognitive" simulation of the work that was analyzed. A wider range of such tools is likely to be available in the future, including some that do a better job of representing work—for example, explicitly representing both conceptual and procedural knowledge.

In the area of speech recognition, machines soon will be able to routinely analyze utterances and, if necessary, translate them into different languages. This capability takes advantage of the constraints placed on language by the specific situations in which it is used, avoiding the general problem of speech understanding, which remains a basic research issue. Nevertheless, instructional designers now need to consider which training or instruction situations need which modalities of interaction. Listening and speaking, while not likely to become the only form of student interaction with training systems, will certainly play a major role in learning environments.

Object-oriented software engineering technology will play a major role in any cost-effective development effort. While such software is clearly needed to develop systems simulations, the new technology will play a big role as well in the design of systems to promote learning. Not only the systems that one is teaching about, but also the very conversations that lead to learning, can best be represented using systems of computational objects. The disciplined use of object-oriented methodology can improve the quality of computer-based simulations and coaches while simultaneously yielding substantial improvements in software development efficiency.

DEVELOPMENT OF TECHNOLOGY FOR
EDUCATION AND TRAINING

Improvements in how software can be developed more efficiently are surely needed when even the largest software house can encounter situations in which a major product is delayed more than a year beyond the expected time of initial release, as often happens today. Thus it is recommended that those undertaking the design of educational software use basic object-oriented techniques for effective tool development and deployment. Furthermore, a modular decomposition strategy will work best—namely, to (1) decompose the task into a number of manageably small pieces; (2) use off-the-shelf capability for as many pieces as possible; and (3) replace any piece-production activity that falls behind with one that is more reliable. Object-oriented design strategies should be used to drive the decomposition process.

The overriding strategy in any such effort should include at least three elements. First, avoid expensive technology except when it is really needed—paper and one-way television can handle many situations. Second, commoditize the tools needed for better learning and make sure that they work on minimal, affordable computer systems. And, third, use coached apprenticeship strategies for educating key workers. Multimedia can be used when direct experience is particularly important.

To end on a practical note, it is always useful to look for ways to combine work and the educational use of computers. Many enterprises use powerful computer systems, but few of these enterprises operate 24 hours a day. Either as part of work time or as an after-hours approach to education, companies could use the same machines that support work during the day for training after hours or even during the course of work. For example, a plant owned by ABB in Helsinki was setting up a new production line to make motor drives. Since virtually each drive was a custom job, computers were placed at each worker's station on the production line to convey parts information and also details on the device assembly. The computers also were equipped to provide additional training. A worker who had questions about how to build a particular motor drive could aim his bar code reader at some special codes on his bench, prompting the computer system to branch to a training package (built in cooperation with the Technical University of Helsinki) that delivered truly "just-in-time" training.

U.S.-based Federal Express uses the computer systems of their office workers to provide training to other employees after hours. While the scheme currently aims only at FedEx workers, a broader possibility, in which the computers at several large businesses in an urban center are used by a virtual junior college at night, seems quite feasible.

The bottom line, then, is that the economic forces that have made information technology cheap and ubiquitous in commerce can and should be put to work for education and training. The actual machines and networks of the commercial sector could be used during the off hours, and the commoditizing effect of industry-driven standardization could be used to cut the costs of learning technology. Finally, the multimedia possibilities of the "edu-tainment" world could be used productively to promote more learning-by-doing throughout developing country educational systems.

NOTES

1. See L. Percival and V. Patel, "Sexual Beliefs and Practices by Women in Urban Zimbabwe: Implications for Health Education," *McGill Journal of Education* 28 (1993); and Sivaramakrishnan and V. Patel, "Relationship between Childhood Diseases and Food Avoidances in Rural South India," *Ecology of Food and Nutrition* (1993): 31.

2. "Commodity" refers to a product that has penetrated the world culture sufficiently to have a large market, multiple suppliers, and thus a low price for both purchase and shipment.

3. Note that within a period of about three years, Bangalore has grown from minimal software industry to being the largest applications software producer region after Silicon Valley!

4. While agreements have placed limitations on publicity about price specifics, it is a fact that both major universities in Pittsburgh can supply such packages to their students for substantially under $100. The arrangements tend to involve network or CD-ROM distribution without printed documentation (it resides instead on the CD or on a network server) and with some investment of technical support labor by the universities.

5. The information reported in this section is from T. D. Tilson, "Sustainability in Four Interactive Radio Projects: Bolivia, Honduras, Lesotho, Papua New Guinea," in *Education Technology: Sustainable and Effective Use*. Document No. PHREE/91/32 (Washington, D.C.: World Bank, 1991).

6. While masculine pronouns are used here for simplicity of discourse, a special value of interactive radio is its utility for female students in cultures where women's access to public places is restricted. Women can gather in a home around the radio and learn together, even helping each other, with the radio teacher's utterances as a prompt.

7. Tilson, "Sustainability."

8. A. Hauptmann et al., "Prototype Reading Coach that Listens: Summary of Project Listen," *Proceedings of the ARPA Workshop on Human Language Technology*, Princeton, N.J., March 1994.

9. Long-distance telephone carriers and some credit card companies already offer voice interactions over phone lines ("If you wish to report a missing credit card, say '1'; for your current balance, say '2' . . .).

10. Kevin Kelly, *Out of Control: The Rise of Neo-biological Civilization* (Waltham, Mass.: Addison-Wesley, 1994).

11. Rodney Brooks, "New Approaches to Robotics," *Science* (1991): 253.

12. Modern software development often consists of a process in which the roles of program components and the world with which a program interacts are designed first. Then each role is captured as an independent piece of program called an object. One goal is to make objects reusable in new programs.

13. A. Lesgold, "An Object-based Situational Approach to Task Analysis," in *Learning Electricity and Electronics with Advanced Educational Technology*, ed. M. Caillot. NATO ASI Series F, Vol. 115. (Berlin: Springer-Verlag, 1993), 291-302.

14. A number of papers address the learning-by-doing approach. One book, accessible only on Internet, is Roger Schank's *Engines for Education* hypertext, which can be accessed through World Wide Web using locator http://www.ils.nwu.edu/ and then selecting the option for the Engines for Education Project. Also see A. M. Lesgold et al., "SHERLOCK: A Coached Practice Environment for an Electronics Troubleshooting Job," in *Computer Assisted Instruction and Intelligent Tutoring Systems: Shared Issues and Complementary Approaches,* ed. J. Larkin and R. Chabay (Hillsdale, N.J.: Lawrence Erlbaum Associates, 1992), 201-238; S. Lajoie and A. Lesgold, "Apprenticeship Training in the Workplace: Computer Coached Practice Environment as a New Form of Apprenticeship," *Machine-Mediated Learning* 3 (1989): 7-28; S. Katz and A. Lesgold, "The Role of the Tutor in Computer-based Collaborative Learning Situations," in *Computers as Cognitive Tools,* ed. S. Lajoie and S. Derry (Hillsdale, N.J.: Lawrence Erlbaum Associates, 1993), 289-317; G. Eggan and A. Lesgold, "Modelling Requirements for Intelligent Training Systems," in *Instructional Models in Computer-based Learning Environments,* ed. S. Dijkstra, H. P. M. Krammer, and J. J. G. van Merrienboer (Berlin: Springer-Verlag, 1992), 97-111; S. Katz et al., "Self-adjusting Curriculum Planning in Sherlock II," in *Lecture Notes in Computer Science: Proceedings of the Fourth International Conference on Computers in Learning* (ICCAL '92) (Berlin: Springer-Verlag, 1992); S. Katz et al., "Modeling the Student in Sherlock II," *Journal of Artificial Intelligence in Education* (Special issue on student modeling, ed. G. McCalla and J. Greer) 3 (1993): 495-518; and A. Lesgold, "Ideas about Feedback and Their Implications for Intelligent Coached Apprenticeship," *Machine-Mediated Learning* 4 (1994): 67-70.

15. The system my group built could be modified to run on a laptop with a built-in CD-ROM.

16. Sivaramakrishnan and Patel, "Relationship."

17. A. Lesgold, "Process Control for Educating a Smart Work Force," in *Linking School to Work: Roles for Standards and Assessment,* ed. L. B. Resnick, J. Wirt, and D. Jenkins (New York: Jossey-Bass, in press).

18. Mosaic is a product of the National Center for Supercomputer Applications at the University of Illinois. It has been licensed to multiple companies, including Mosaic Communications Corporation and Spry, Inc. Mosaic is a "reader" that permits access to hypermedia information servers using a convention called hypertext markup language (HTML), developed initially at the European Organization for Nuclear Research (CERN) in Switzerland. HTML documents are distributed, in the sense that a word or image in one document can point to an information file in another document anywhere on the Internet. The collection of HTML servers worldwide is known as World Wide Web.

Technological Innovation and Services

JORDAN J. BARUCH
Jordan J. Baruch Associates

The growth in recent years of services as a component of national economies has drawn more than a passing glance from the World Bank and other institutions concerned with Third World development. Indeed, the extent of this interest in services is evident in a 1994 Bank publication, *Liberalizing International Transactions in Services: A Handbook.* In the services area, technological changes are producing impacts as important as those they have produced in mining, manufacturing, and agriculture. Not only are the changes important, but they also are blurring (if not obliterating) the distinctions between services and the other economic sectors.

James Brian Quinn et al. have pointed out that today in the developed nations, services are large, technology- and capital-intensive industries that are virtually inseparable from manufacturing.[1] Elsewhere, the National Research Council has described information technology and its impact on the performance of services in developed countries.[2] In looking at services, these publications have analyzed, in terms relevant to the Bank's mission, the impacts of the traditional service sectors—finance, real estate, insurance, retail and wholesale trade, transportation, communications, and sometimes construction—under the forces of technological change, on the overall economy, including manufacturing. While these analyses have focused on the developed world, there is little reason to expect significant differences in the developing countries.

But rather than review the implications for the developing countries of what already has been written, this paper will focus on the technologies of information-based services and their implications for developing countries. The first section, on *embodied services* (a term borrowed from the concept of embodied technologies), focuses on the transborder trade in goods that are, in essence, indistinguishable from services. The second section, which addresses transborder services, focuses on information-based services in the developing countries that serve both as import and export activities.

While there may be many questions about the predictions made at the Symposium on Marshaling Technology for Development, one thing is clear: the growing technological, managerial, and political sophistication of the developing countries will produce a strong force for the vertical integration of their industries and their services. Indeed, they will exploit every opportunity to secure added value by moving up the input scale while developing capital goods and using them to fill in the lower levels. Does this sound familiar?

EMBODIED SERVICES

Some important technological changes taking place in the Third World depend heavily on information technology, but they have become evident in more material ways. For example, a women's clothing company called The Limited gathers copious data at its point-of-sale terminals which are analyzed almost continuously at its central office. Based on these analyses, the firm predicts when it must reorder a particular item. The transaction order is then automatically generated and, once approved, is transmitted electronically to China or wherever that item is to be fabricated. When the order is received, it is filled and the garments are shipped by air back to drop points in the United States. Thus within days of the order, the goods are in The Limited's retail outlets.

Certainly, the data gathering, computing, transmission, transportation, distribution, and retailing are services. But how about the actual fabrication of the garments? Under the influence of today's drive by the industrialized countries to outsource anything in which a firm does not excel,[3] fabrication also must be considered a service. A closer look at the potential positive and negative impacts of technology on services in the developing countries may reveal that this kind of fabrication is one of the more important services.

Research and development services performed in the industrial nations may, under the right conditions, be particularly amenable to embodiment in goods generated in developing countries. Such a transfer has been seen in electronic fabrication, agriculture, and food processing, but it is far from simple. For example, recent discussions among the National Mariculture Center in Israel, the government of Jordan, and representatives of the World Bank have highlighted an interesting question that is likely to have ongoing implications for services in the developing countries. In Elat, Israel, adjacent to Aqaba, Jordan, the National Mariculture Center has developed exquisite technologies for the genetic manipulation, culture, and care of several fish species that are in great demand. For the gilt-head bream, an expensive saltwater fish marketed in France as dorade, the center has succeeded in producing a continual flow of bream fingerlings throughout the year. Jordanians recently expressed a strong interest in forming some kind of partnership to share in this activity, although the two countries differ widely in their levels of technological sophistication in this field. Yet conversion of the fingerlings into marketable, edible fish requires a large land area for ponds and access to saltwater—both of which Jordan has. It also requires careful feeding, control of disease, and good harvesting practices—skills that Israel knows how to teach. The conversion of fingerlings to fish is a kind of farming. Indeed, the Israelis now doing the conversion are not highly trained technologists or skilled biologists or ichthyologists; they are farmers. They, in fact, are the service branch of a partnership that converts the center's R&D output into an economic product. And clearly these fish farmers could work in Jordan or Egypt as easily as in Israel.

While the technical barriers to a partnership between Israel and Jordan are small, the organizational ones require attention. For example, such partnerships help to shrink the developmental gap between the developing country and the industrialized partner only if both countries benefit. Thus the flow of skills to the developing countries must be encouraged, but the economic position of the more developed partner must not suffer. Under this paradigm, the Jordanians might well start out as fish farmers, but it would be in the partners' joint interest for the Jordanians to move up the technological scale. They would learn to culture, rather than import, *Rotifera* for feeding the young fish. They also would learn to compound fish food for the mature fish, induce spawning, inoculate fish, and gradually take over the whole gilt-head bream culture process. The Israeli center would share in the profits and remain available for consultation and process improvement. It also would extend the process to other fish such as the grey mullet, which is in demand in the Arab world. The partners, to both of their benefits, might well choose to extend their partnership to include Egypt. There, a copious supply of fresh water would permit the raising of various freshwater species, including ornamentals.

Such two-level partnerships are possible in other areas as well. Some organizations in the developed countries will find it profitable to share their design and marketing skills with partners in the developing countries. Why partnerships? The fact is that the developing countries will tend increasingly to use their scarce natural resources to secure a developmental advantage in bilateral agreements. For example, in Indonesia or Thailand skilled contract furniture makers with access to exotic trees might set up two-level partnerships with developed country design and marketing firms. If access to those trees is restricted (by law or pricing), such partnerships can become a significant force in the furniture field and advance the technological level of the junior country. Similar potentials exist in the field of pharmaceuticals, where developed country firms are constantly seeking botanicals that can form the basis for new drugs. To this end, skilled botanical observers could collect and tabulate the thousands of individual species in untouched forests in the developing countries. Once these samples have been identified and their potential verified by scientists and technologists in the developed world, developing countries could assume the task of farming and even converting those natural materials. The "upper-level" partner would, of course, be responsible for proof of effectiveness, regulatory compliance, and marketing.

The growth of such partnerships will require the establishment of legal bodies and agreements that can enforceably adjudicate disputes between such partners and between them and their clients. That, however, must be the subject of a later discussion, but one in which the World Bank could play an important role.

TRANSBORDER SERVICES: DATA EXPORTING

The very concept of transborder services is a child of modern technology. Today, however, communication technology, computer technology, and dynamic entrepreneurship have combined to make certain transborder services barely noticeable within the developed nations they serve. Consider the following predictions:

1. The cost of digital electronic communication is becoming independent of distance, and its price will, because of growing competition, drop accordingly.

2. The cost of electronic communication bandwidth is dropping faster than a factor of two per year and its price will follow.

3. The price of digital memory of all kinds is dropping even faster than that of bandwidth.

4. Manufacturers, including software creators, will continue to see the displacement of service providers by products as a profitable strategy.

The massive drops in communication costs predicted above will have major implications for developing countries. The plummeting prices for both distance and bandwidth in communication would seem to foretell a rosy future for the growth of transborder services. Such long-distance services as data entry and 800 number response already have grown, and, as the costs of computation and communication decrease, it seems reasonable to expect that those kinds of services will grow and flourish. Indeed, in some developing countries the costs of communication services and locally networked desktop computers have become low enough to permit the growth of communication or data-processing zones, where moderately trained and educated service providers perform such data services far from the source of the raw material. When services once performed in developed countries are shifted to the developing countries, they become a significant source of both income and technological familiarization. But the same kind of technological developments that have made those services possible will act to change them in the future.

Prediction 4—the urge to replace service providers with "things" that can do the same job—implies, however, that growth is only part of the story. Consider just two recent trends and events: scanner technology and voice-input technology are improving rapidly, and the Microsoft Corporation is in the process of absorbing Intuit. What do those developments have to do with long-distance services? The acquisition of Intuit by Microsoft signals a major potential change in the check- and credit slip-clearing processes. Intuit currently offers its subscribers, as part of its accounting package (over 5 million in use), a program element called Billpayer. For 15 cents (less than half the price of a first-class stamp) the user can enter a transaction and have it paid automatically. Because there is no check, no clearance (and thus no overseas clearance service) is needed. Intuit also offers a register system for recording credit purchases. Soon that register may be checked electronically (if not entered directly) against the card issuer's records, eliminating the need for overseas credit-slip transcription. These developments are not unlike the shift made earlier in this century by the telephone system from operators to dials. In both cases, hardware and software products were used to shift the provision of services from some outside provider to the user.

But this process will not stop there. Customers in the industrialized countries will increasingly use the digital dial network along with scanners and voice recognition systems for entering all kinds of data now being processed by people here and abroad. As scanners become able to read less-disciplined inputs, and when the corrections to misinterpretation can be made by voice, the need for a large labor force to interpret written documents will be markedly reduced. In light of these developments, it can be reasonably expected that the use of developing country data-entry facilities will grow as the price of long-distance communication drops, but then will shrink rapidly as the need for that service declines, much like the development of the home permanent displaced many hairdressers (service providers) and the advent of permanent-press fabrics, along with the development of new washers, dryers, and detergents, eliminated whole sections of the commercial laundry business.

What, then, will happen in developing countries with the demise of communication zones? What will replace them? One possibility may stem from the impact of catalog shopping in the United States, where it has been an important development in retail sales. On-line catalogs are starting to make their appearance on the Internet, and the televised Home Shopping Network has been valued at over a billion dollars. It is reasonable to predict, then, that a direct product export business will grow in developing countries, where, at the moment, some of the enormous variety of crafts available in the local markets—carved wooden flowers from Bali, gold jewelry from India, sculptures, blankets, and even furniture—is sold abroad through multiple transactions and multiple price hikes to the eventual consumer. Skilled entrepreneurs in the developing countries will form local cooperatives, corporations, kibbutzim, or centers that will prepare and transmit on-line catalogs of local products directly to potential customers in the industrialized world. They, in turn, will then place their orders from home using secure credit card readers. Fund transfers will take place electronically as will customs collections (if any).

But how will U.S. businesses respond? It seems clear that domestic catalog dealers will try to compete by outsourcing as much of their business as possible. They may start by sending detailed manufacturing specifications to the developing world (as does The Limited), but soon they will realize that the added value to their product lines is consisting primarily of quality control. If they can arrange to transfer that function to the developing country, the goods can be drop-shipped. New overseas shipping arrangements probably will be created to meet the demands for rapid delivery from single overseas points to multiple consumer locations in the developed countries.

The dramatic, ongoing drop in local information storage costs will continue to stimulate new developments, some of which also will have significant implications for the developing countries. The actual nature of those developments is hard to predict because they depend so heavily on user preferences. As an example, for many information applications, memory (prediction 3) and bandwidth (prediction 2) are interchangeable. Economically, whether data are stored locally or remotely depends in large measure on the relative cost of memory and data transfer. But despite that interchangeability, many users prefer the privacy of local memory to central storage (just as telephone customers continue to prefer local telephone answering machines over central answering services).

In developing countries, literacy is currently a significant barrier to the use of such stored communication modes as electronic mail, facsimile transmission, and even written instructions. Yet stored (time-shifted) communications are essential for instructions, record keeping, and general reference. In the near future, at least, illiteracy will continue to pose problems, especially among the adult population, but because of both the demand and the technology, "voice fax," in which the voice provides both the address and the data to be transmitted, will grow to fill that gulf. And illiterates will not be the only ones to benefit. Any telephone, wired or cellular, will then become a "voice fax" machine, and stored communication, because of its new convenience, will spread rapidly and significantly among the mobile and isolated members of developing country populations.

One significant form of stored-data communication presently used in developed countries is the electronic bulletin board, in which users post inquiries and other users post responses. While many bulletin boards are quite informal (such as computer user groups), others are more structured, even to the point of identifying all respondents to provide for the authentication of their responses. The growth of that technology and the continuing drop in memory and bandwidth prices offer the opportunity for the development of a new aspect of transborder information services.

TRANSBORDER SERVICES: INFORMATION IMPORTING

While the electronic export of data services is profitable to the developing countries now and will be in the foreseeable future, the electronic import of information will be far more important. The critical factor leading to the potential for importing information is the explosive growth and availability of the Internet. From 1984 to 1994, the number of computers linked via the Internet grew from just over 1,000 to over 3.8 million. Consider what an inquiry service on the Internet could do for the Third World. Questions of the "How do I . . . ?" type could originate in the developing countries and be answered anywhere in the world by volunteer organizations such as the International Executive Service Corps, professional consultants, or others. The answers might start as simple text but would soon range from drawings through voice messages to full-scale video demonstrations. Yet a one-question/one-answer system would represent a terrible waste of intellectual effort and knowledge. The answers must be useful to others besides the original inquirer and their persistence in time must be ensured.

This could be accomplished by the creation of a Third World application-oriented inquiry and library system on the Internet, described in Figure 1. The subscripts included in the figure indicate that users, inquiries, experts, and so forth, are all identified so that they can be tied together for future reference and quality control. Note that the user not only gets his or her inquiry answered but has an opportunity to evaluate the response. "Did it work?" "What were its limitations?" "What other questions does it raise?"—iterative query. Indexers note the application to which the responses are relevant, but they do more than simply index the response by the one application. Chosen for their experience, they are able to envision other potentially relevant applications. A question about distributing water to a row crop, for example, may well generate responses applicable to drinking-water distribution or evaporative cooling. Without application indexing, the direct and extended intellectual links would be lost. Subsequent inquiries about those applications would be unable to draw on what the system already "knows."

Unlike the inquiry-driven system described in Figure 1, a publication-driven system is featured in Figure 2. Both systems, however, are aimed at making world knowledge easy to access and use in developing countries. Published scientific and technical papers are generally classified by the technology with which they deal, and authors and publishers alike are, rightly, concerned with making the papers accessible to the authors' peers and others who will build on the new knowledge contained in them. Eventually, the knowledge embodied in such papers is put to use—often in fields far removed from those of either the publishers or the original authors. If those users also publish, citation trails can lead back to the original articles. But even that publication process is most likely to occur in the developed world.

Figure 2 depicts a situation in which two people—a user and a special kind of reader, an annotator—interact with the information in the published paper. Both participate, separately or in tandem, in creating "electronic marginalia" describing an application to which the information has been, or might be, put to use in the Third World. Initially both the user and the annotator may be in a developed

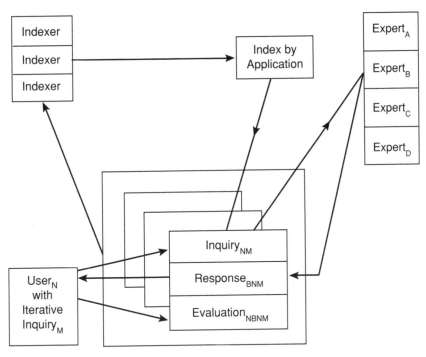

FIGURE 1   Application inquiry, response, and index system.

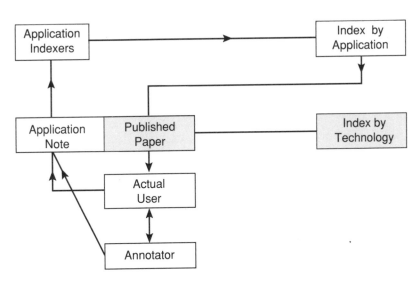

FIGURE 2   Application annotation and indexing system for published material.

country, but with time, as was the case in the inquiry-driven system, Third World users will add their experience. The annotated document can now be indexed as a single unit.

At this point the two systems come together. The application notes of Figure 2 and the evaluated responses of Figure 1 are directly useful material. Via the application-indexing process, they will be listed in a single index. Got a problem? Welcome to one-stop answer shopping!

Depending on the degree of use of the system and on demand, the material can be gathered by application for wider distribution via compact disks (CDs) or other means. Such off-line use can save communication costs for distant users and provide services to isolated ones. The resulting collection, regardless of its form, will have been stored, catalogued by application, and organized into the kind of user-focused technological library that will accelerate the development of any developing nation that has access to it.

CONCLUSION

Any other author asked to deal with the subject of technological innovation and services might well have adopted a wider perspective than the one used here. But the organizers of this symposium have asked, "How can technology be marshaled for development?" This attempt to answer that question has come back to information services—and this is no accident. Indeed, one might say that the rule of the future will be: among the so-called services, a country's ability to manage and use information will be the greatest single determinant of its rate of development.

NOTES

1. James Brian Quinn, Penny C. Paquette, and Jordan J. Baruch, "Exploiting the Manufacturing Services Interface," *Sloan Management Review* (summer 1988).

2. National Research Council, *Information Technology in the Service Society: A Twenty-First Century Lever* (Washington, D.C.: National Academy Press, 1994).

3. See, for example, James Brian Quinn, *Intelligent Enterprise* (New York: Free Press, 1992), and the end notes for chaps. 2-4.

Health Technology and Developing Countries: Dilemmas and Applications

KENNETH I. SHINE, M.D.
President, Institute of Medicine

Three caveats, or principles, will preface this look at health technology and developing countries. The first caveat, which distinguishes the health sector from all others, is this sector's special character: the status of human health derives from the dynamics and outputs of every other aspect of society to a unique extent. Not only the level but also the equity of distribution of a society's economic product determine, though not exclusively, health status. This economic product includes the availability of infrastructure—clean water and air, safe work environments, transportation networks to link people and communities to health facilities, communications media to link them to health information, and health facilities themselves. The education sector has a relentless and unequivocal effect on health status. It provides for the education of women, now widely accepted as crucial in child and family health and welfare; the education of health professionals who know not only how to use health technology but also how to care for people; and the development of a critical mass of individuals who can perform research and analysis in all the areas that pertain to keeping people healthy and making them well. All this being said, health status and economic status and educational status and good roads are not inevitably correlated. Moreover, the poor, the uneducated, and the politically powerless bear a much larger burden of death and disability than their opposites—the wealthier, the more educated, and the enfranchised.

The second caveat, or principle, is a converse of the first: health status not only derives from every other aspect of society but also contributes to it. This is so despite the fact that economists like to relegate health and education to the so-called social sector—that is, the sector that costs money but produces nothing, unlike agriculture and industry which are the bulwarks of the productive sector. This, however, is nonsense. For example, children who grow up in toxic environments are less likely to prosper, educationally or economically; they are, therefore, less likely to be productive in societal terms. Women who are in poor health are not able to take care of their families as well as healthy women do. They also tend to have infants who are sicker and who die earlier. As a result, they have more babies to replace those who have been lost. The costs of the consequent high morbidity and high fertility are not trivial in development terms.

The third caveat is that, precisely because of the intimate and extensive links among health status and every other aspect of human existence, technology is only part of the answer. Very high technology is an even smaller part, whether talking about developed or developing countries. In fact, in itself technology offers little in the way of solutions to many of the factors that affect human health and well-being negatively—smoking and other substance abuse, lack of exercise, ruinous diets, sexual carelessness, noncompliance with medical regimens, or wearing a seat belt or fixing the exhaust on a vehicle are matters of behavior. And, other than the case of firearms, where *removal* of technology could do a lot for human health, there are no technological answers to the problems and consequences of all kinds of violence.

CHARACTERISTICS AND TRENDS IN HEALTH AND MEDICAL TECHNOLOGY

These principles having been established, what then are some of the characteristics and trends in health and medical technology, generically and as they apply in the United States?

Technology and Fundamental Research

The United States plays a commanding role in the development of health and medical technology, in part because most European and Japanese as well as American manufacturers earn more than half of their profits from U.S. sales. Thus even research not performed in the United States is influenced heavily by U.S. requirements. The main reason, however, is the U.S. leadership role in basic research, or what is increasingly referred to as *fundamental science*. The term is an excellent one because it diverges from the stereotypical view of basic research as remote and somehow irrelevant to daily life in the "real" world. To the contrary, *fundamental science* emphasizes that the concept of science—in this case, medical science—is an integral part of an entire research and development process that, in theory, leads to better health. By extension, the term suggests that the way to start this process and achieve technological success—particularly in the life sciences—is to foster new and creative ideas in fundamental science laboratories.

A good example of how "fundamental" becomes "applied" and, therefore, "relevant" is the classic, seminal work of Stanley Cohen and Robert Boyer in bacteria that opened the whole new world of biotechnology. Other areas in which industrial frontiers were opened by fundamental science were the research underlying cardiopulmonary bypass surgery and the characterization of the metabolic pathways for insulin regulation and cholesterol metabolism, which spawned another generation of pharmaceuticals, as well as fresh understandings of prevention and clinical management. The open transmission of scientific findings, the ease of communication between industry and fundamental scientists, the availability of capital, and, not least, the reward system in the United States, combined to fuel these development processes.

But such processes are not orderly. Health science technology is notable for not following the neat sequence of steps in technology development that often characterizes other fields. The general notion of some kind of fundamental scientific discovery that first informs a body of applied science, which then translates it into technology, which then is properly engineered and manufactured for application to patients, is just plain inaccurate for the health sciences. In the "real life" of medicine, it is not uncommon for basic scientists to identify a gene, construct a gene product, file a patent, obtain an approval for clinical trials, and rapidly characterize a new therapy. Somewhere in the process, there is involvement with the Food and Drug Administration (FDA),[1] but that involvement does not always occur at the same point or in the same way. And somewhere in that process a new company may be formed, a pharmaceutical house may enter into a collaboration or even fund some of the trials of the item in question, and, with a demonstration of efficacy through a licensing mechanism or some other strategy, a new product may be born.

Or if the environment is right, the usual steps from fundamental science to commercial application may be short-circuited by being carried out by a single researcher or company. One example is the history of the use of fiber-optic techniques to peer into internal organs and blood vessels. In what might be called "unilateral serendipity," a single gastrointestinal physiologist, in pursuit of answers about acid secretion in the stomach, designed (with a graduate student) a way to coat optical fibers. This led to the gastroscope. A postscript to this brief history illuminates the complexity and the occasional irony of what happens in U.S. R&D processes: although the technology was created in the United States, the manufacture of fiber-optic instruments was subsequently perfected by Japanese companies, and Japan now largely controls the world's market for such tools.

The Health Technologies Market and Its Implications

The criteria for the approval of instruments, drugs, and treatments in the United States are limited to efficacy and safety. There is absolutely no requirement to show cost-effectiveness or improved productivity. The market and health implications of this aspect of the approval process are enormous. In an environment in which technology is applied by physicians and paid for by third parties, there is little role for consumer (patient) choice. In other words, the large universe of patients really has little to do with the diffusion and rate of application of health technologies; these factors are controlled by the "prescribers" of medical technology and their "reimbursers."

It is not surprising that an imperfect, "un-free" market in which the consumer carries so little weight would become very high priced and, consequently, would play what many consider to be an inordinately large role in overall health sector costs—costs that in 1995 will approach $1 trillion annually. Indeed, between 25 and 50 percent of the rate of increase in health care costs in the United States stems from the way in which technology is diffused. The costs of technology diffusion include not only those for production, distribution, and marketing, but also the amount that the producing company calculates it has spent on research and development, as well as what it calculates provisionally for the costs of liability. And then there is profit: for most major European, Japanese, and American manufacturers, one-half or more of their net profits comes from sales of medical technology in the United States alone. By extrapolation, each of the components of health technology economics has a big dollar amount attached to it. For the R&D component, one current estimate is that the U.S. national investment in health and medical research is about $25 billion.[2]

In macroeconomic terms, the amount of growth in health sector technology is gratifying. Given the figures alone, it is not surprising that the United States has been the driving force behind technology development worldwide. But these figures are not necessarily gratifying in human terms: there is relatively thin evidence that this technological burgeoning has, in fact, improved the health of the population as a whole.

Technological Disappointments and the Dilemma of "Social Products"

Will the developing world benefit from what is going on technologically in the United States? Despite the enormous size of the investment in and the indisputably high caliber of the U.S. health and medical technology subsector, it gets mediocre marks in terms of generating new products that are critically relevant to the needs of the developing world. Vaccines are a good example. While it is true that success in immunizing the world's children over the past decade has been dramatic, this success has not stemmed from any new, quantum leap in technology but from the achievements of the Expanded Program on Immunization (EPI), which is administered by the World Health Organization. Yet despite EPI, 20 percent of the world's children remain unvaccinated and immunization gains are slipping in a number of places, including the United States. Better childhood vaccines are needed everywhere. Ideally, such vaccines would be given near birth as a single dose, would contain multiple antigens and protect against diseases not currently targeted, and would be heat stable and affordable.

But development of such a "dream vaccine" is likely to take a long, long time, not just because the technology is complex but also because presently the commercial aspects of such a product do not seem enticing to industry. With the cost of new drug development in the range of $200-$250 million, the financial motivation for pharmaceutical companies to make substantial investments in development of a vaccine for which a consumer requires only one, two, or three doses is limited. Along the same lines, a consortium of American pharmaceutical companies recently decided to jointly develop antiviral agents against the human immunodeficiency virus (HIV). Such therapeutic interventions characteristically require repeated doses over a prolonged period, entailing greater sales volumes and thus more profit. This reveals clearly why a joint effort to develop an AIDS vaccine was not the focus of such a consortium.

A related example is the development of contraceptive technologies. There is growing consensus that the menu of options available to individuals and couples for planning the spacing and number of their children is deficient. Each method presently available has limitations; only one male contraceptive is reversible; only one contraceptive protects against both impregnation and infection; and most methods have side effects for at least some women—women everywhere, not just in the developing world.

Based on recognition of this state of affairs and wider recognition of the intimate dynamics among population growth, the environment, and poverty,[3] a consortium of funders asked the Institute of Medicine (IOM) to explore what new leads emerging from contemporary science would offset the various disincentives associated with such contraceptive products and would attract young scientists and motivate industry to come back into the field.[4] Among the disincentives are the complexity of the science itself, the costs of R&D, issues of liability, and a cluster of political considerations. In the IOM study, one area of focus was the possible development of a contraceptive vaccine.

One of the great misconceptions of policy makers is that fundamental scientists are motivated primarily by curiosity and a passion for science. While scientists surely are driven by both, the motivation for an individual basic scientist to dedicate substantial effort to the development of any kind of new and novel vaccine seems constrained. The efforts of scientists are determined by the problems perceived as commanding within their fields *and* by the likelihood of significant other rewards.

In summary, the United States has produced a vast amount of very high technology, yet the economic costs of that progress have been great and the payoff in terms of overall, improved health status has not been commensurate. Over the last two years, a great deal of thought has been dedicated in the United States to reshaping the role of medical technology, as well as the physical infrastructure in which it resides, so that they are less financially burdensome and better adapted to the kind of health care system to which Americans might well aspire. The fact that U.S. health reform is on hold does not mean that these thought processes have been or should be arrested; they will continue and will be tested, especially at the state and local levels. In this connection, great care must be taken not to seduce the developing countries, particularly through misguided "sharing" of technology, into making precisely those mistakes Americans are trying, with great difficulty, to correct—another caveat.

HEALTH CARE AND TECHNOLOGY

A number of new medical technologies and approaches to health care delivery are likely to have significance for all countries, developed and developing.

Strategy Shifts: The Expansion of Outpatient Approaches

The proliferation of diagnostic and therapeutic technologies is allowing a growing proportion of health care to be provided on an outpatient or ambulatory basis. Even such demanding procedures as cardiac catheterizations, as well as a rising number of surgical procedures, are increasingly being performed in outpatient settings. Furthermore, the average lengths of hospital stay are falling steadily. Even in institutions that perform the very major surgeries (which drive up averages)—for example, organ transplants—average hospital stays are less than six and a half days. In many tertiary care institutions the average length of stay is under five days. As for specific procedures, 10 years ago observers were impressed by the capacity of the medical community to perform vasectomies and tubal ligations on an outpatient basis; five years ago they were impressed by the capacity to perform breast biopsies and plastic surgeries on an outpatient basis. Now it is feasible to remove a gall bladder using a fiber-optic technique with, at most, an overnight stay. In the near future, no hospital stay whatsoever will be required.

U.S. hospitals, then, are eliminating beds and building ambulatory facilities, "surgi-centers," and satellite clinics. A recent review of the intramural programs at the National Institutes of Health (NIH) recommended that the capacity of the clinical center at this highly research-intensive institution be reduced by 50 percent. More and more centers are building hotel and motel accommodations so that patients from outlying areas can stay nearby with their families for a day or two while undergoing ambulatory procedures. This quite massive change not only is a function of changing technology but also is intimately connected with the urgent need to control health costs. Furthermore, staying out of the hospital has other benefits for both human and financial health. The potential for nosocomial infections is dramatically reduced in outpatient settings, as are many of the other complications that can result from inpatient treatment and case management. All of this, of course, saves money: the costs of nosocomial infections were recently estimated to be $5-10 billion annually.

Thus in the near term, in developed and developing countries alike, the most appropriate health care delivery model will be ambulatory diagnostic and procedure rooms with nearby, low-cost hotel accommodations and a relatively small number of very high-tech hospital beds. Investments in ambulatory services, together with an emphasis on comprehensive primary and preventive care, clearly will be the best investments in health care in all parts of the world.

Information

In an era of cost containment, information is everything. One of the most important types of information is data on outcomes of encounters with the health care system, produced by so-called medical effectiveness research. This research encompasses existing clinical practices, the development of practice guidelines, technology assessment, and cost-effectiveness analysis. It also includes research on the effectiveness of health promotion and disease prevention programs and interventions.

All of these areas are highly relevant to the focus of this symposium and the work of the World Bank. The Bank already has moved in significant ways in this field of inquiry through the preparation of the *World Development Report 1993: Investing in Health*.[5] The Bank's development of the disability-adjusted life year (DALY) offers a quantitative approach that permits researchers to calculate the burden of disease in a given societal setting by measuring the number of disability-adjusted life years produced by a certain illness, and permits policy makers to make those investments that, for a given expenditure, will maximally reduce the number of DALYs in that setting.

Assessment of the outcomes of technological applications are intensely pertinent to these calculations. In considering the potential power of such assessments, one need go no farther than vaccines, which are among the most cost-effective interventions available. A 1985 Institute of Medicine study[6] demonstrated that, in the United States, a dollar spent on vaccine development saved 10 dollars in health care costs; that ratio is now thought to be about 1:12. The 1985 calculation did not include the accrued long-term benefits to patients and the associated pro-rated savings. Recent work on the impacts of infectious diseases in adults in developing countries suggests that ratios in those contexts might be even more favorable.[7]

Clearly, then, health care systems that provide outcome and effectiveness information are of utmost importance to any government. They already are pivotal in the United States where, increasingly, the rising number of managed care organizations must justify expenditures with good outcomes data that enable planners and managers to draw conclusions about the merits of technologies applied, quality of care, effectiveness of case management alternatives, and the relative cost-effectiveness of each.

But the U.S. health care community does not have all the answers. While health services research and technology assessment are a rapidly growing area in the United States and in Europe, there is much more work to be done. A recent congressionally mandated study by the U.S. Office of Technology Assessment[8] concluded that the U.S. federal government's efforts in medical effectiveness research have fallen short of expectations. The reasons include excessive optimism, insufficient funding, and a timid mandate for the lead implementing organization, the Agency for Health Care Policy and Research (AHCPR). One conclusion of the OTA study relevant to the international concerns of this symposium is that large administrative databases, which were the cornerstone of AHCPR's Patient Outcome Research Team (PORT) program, have not proven to be as useful as was hoped in answering questions about the comparative effectiveness of medical treatments. The report notes that, despite the fact that they are potentially powerful sources of information, prospective comparative studies, particularly randomized controlled trials, have been underused. The report suggests that investment in community-based research infrastructure for clinical trials might be well placed.

Behavioral Research and Nontechnological Challenges

Health services research encompasses the way in which health care, health promotion, and disease prevention services are provided at the clinic level and in the community. Within health services research is behavioral research, a growing area of inquiry but a very difficult one. Yet there is substantial pressure to understand better the factors that influence individual and community behavior and how to encourage healthy decisions about behavior—about smoking, drug and alcohol abuse, diet, and exercise.

Another challenge is violence, not only as a matter for law enforcement but also as a public health problem and as a focus for science. As scientific debates go, the war of words over what has been called "the genetics of violence" has itself been marked by violence. This "violence about violence" is a function of frustration about a lack of remedies: law enforcement is daunted; social strategies have met only limited and erratic success; and science and technology have produced no cures. A recent conference sponsored by the National Research Council and Harvard University's John F. Kennedy School of Government concluded that, under the prevailing circumstances, "prudent public officials must respond to violence more like medical researchers following promising leads in a search for a cure than like physicians confidently prescribing a proven therapy."[9] The price of present inabilities is huge: over $450 billion annually in the United States in direct costs and such indirect costs as the loss of economic activity in high-crime areas.[10] The "referred" costs in human distress are not included.

Another daunting and partially related area is biobehavioral medicine and mental disorder. Knowledge of these subjects in the developing world always has been very scattered and weak, but this situation should be substantially remedied by the imminent publication of a study that is the first to explore systematically what is known about mental health, neurologic disorders, and behavioral problems in developing countries.[11] This study's premise is that, as infant mortality continues to decline and life expectancy grows longer in developing countries, chronic illnesses will exert the most pressure on the health systems of those countries. Already half the burden of disease in the developing world stems from injuries, unintentional and intentional, and from noncommunicable diseases. Of the latter, the largest category is "cardiovascular diseases," closely followed by "neuropsychiatric illness." The total mortality and morbidity figures attributable to some composite of neurological and behavioral disorders, intentional injury, internal civil strife, war, and refugee status are presently incalculable. Indeed, the costs of such conditions, for individuals and communities and for developing country economies and health systems, are simply unimaginable.[12]

The Genetic Revolution in Health and Its Implications

The genetic revolution in health offers glorious prospects for human health—as well as not-so-glorious ethical and practical dilemmas, and substantial costs. Eventually it will be possible to eliminate certain diseases at the gene level—that is, at the source rather than when they manifest as illness and symptoms to be treated or palliated. But, although nearly 100 gene therapy experiments have been approved by the U.S. government, they are still just experiments. Researchers have been humbled over the past few years by the complexities of converting knowledge about genes and disease into practical solutions for people who are sick. Thus despite progress, genetic treatment has a long way to go and is not likely to become a standard of care anywhere for many years, allowing time to consider its implications thoroughly.

Genetic screening—the capacity to identify an individual's genetic predisposition to illness—will be a ready tool by the end of this decade. Screening is especially promising in situations in which therapies or mechanisms are available that may postpone onset or prevent "disease X" in individuals predisposed to develop it. At the same time, there are risks and costs. First, the initial impact of screening will be an escalation in health care costs because every positive test will require confirmation and reconfirmation. In addition, prevention or containment of some diseases may require lifelong treatment and monitoring. Second, extraordinary moral and ethical issues are related to how much individuals should or would want to know about their genetic future. Carrying out screening without extensive genetic counseling would be irresponsible. Third, there is heated debate within the U.S. medical scientific establishment about whether any genetic information should be provided to individuals, absent the availability of any clear and beneficial response to that information. Finally, the impact of genetic screening on abortion rates could be profound, and there are obvious implications that fall under the heading of eugenics.

TECHNOLOGY NEEDS OF DEVELOPING COUNTRIES

The development of seven technologies or approaches would meet the urgent needs of developing countries, although priorities among the following technologies can be argued and, indeed, will vary from among and within countries and regions:

1. New reproductive health technologies, especially technologies for male and female contraception
2. Cost-effective approaches to making crucial micronutrients available to populations requiring them
3. Vaccines that protect more efficiently against the usual childhood diseases, as well as against diseases that carry large, unaddressed burdens of morbidity and mortality
4. Expanded and strengthened capacities for the provision of primary care and outpatient clinical care, including the use of cost-effective diagnostics and appropriate therapeutic responses to the diseases identified
5. Cost-effective interventions to prevent and manage the growing prevalence of chronic illness—heart disease, cancer, stroke, lung disease, and diabetes
6. Pharmaceutical and information technologies. Pharmaceutical technologies should anticipate or compensate for problems of drug resistance, and information technologies would make possible the early identification of emerging infections.
7. Protection against infection by the human immunodeficiency virus (HIV).

New Reproductive Health Technologies

The term *reproductive health technologies* is used to broaden the concepts of contraception. Because the contraceptive methods presently available are all limited in some aspect, the principal objective of trying to reinvigorate research and development in this area is to expand the array of good technological options available to couples and individuals in different situations in different cultures at different points in their life cycles. This broader term also embraces an utterly compelling scientific requirement: the need for technologies that protect against infection, whether or not they protect against conception. The over 50 classic sexually-transmitted diseases (CSTD) presently identified produce an enormous burden of morbidity and, in some cases, mortality. To these must be added the acquired immune deficiency syndrome (AIDS) and its causative agent, HIV, and its inevitable mortality.

For most populations of the developing world, the greatest burden of CSTD (with the exception of syphilis) and HIV/AIDS falls on females. Because women are the reproducers, in contrast to the male role of progenitor, their reproductive health is directly linked to the health of their offspring, with the consequent epidemiological and public health impact. Thus, while there is a compelling need for contraceptives for use by males so that they can share the responsibility for conception, there is perhaps an even more urgent need for methods that are controlled not by males or by health care providers but by women themselves.

Finally, whatever the political dimensions of the issue, it is a social and medical fact that in the armamentarium of reproductive health technologies there is a transcendent need for postcoital agents, particularly for situations in which intercourse is relatively infrequent and unpredictable.

The question of whether the biotechnology and large pharmaceutical industries, and ultimately the private sector investment community, can be motivated to engage in this area of research and development remains moot. However that evolves, the demand for capital at any point in the R&D trajectory will remain high.

Micronutrient Research and
Technology Development

Micronutrient deficiencies appear to be widespread in the developing world, with the impact of certain deficiencies (such as vitamin A) extremely large and meaningful in developmental terms. It is feasible to redress these deficits, but in some cases it is not yet technologically straightforward. Some interactions among micronutrients and between micronutrients and certain toxicants (such as lead) are synergistic in ways that might be deleterious, but they remain poorly understood.[13] For example, given the fairly clear causal relationship between high blood-lead levels and cognitive impairment, the insult of lead exposure for an iron-deficient child might be significantly greater than in a child not deficient in that particular micronutrient. Similar questions have been raised in connection with calcium and zinc, but they remain unresolved. Science may have to ask these potentially large questions before technology can be developed or appropriately applied. Science also may have to question the trade-offs between the applications of pesticides, fungicides, and fertilizers and risks to human health in developing countries.

With the exception of vitamin A technologies, which have had some success, the transfer of technologies that could redress other micronutrient deficits at the country level seems to be somewhat turgid and certainly uneven.

Vaccine Development

Lewis Thomas has described the immunization process as one of the genuinely decisive technologies of modern medicine.[14] Indeed, it is highly cost-effective in public health terms, but it has little appeal in the domain of commercial research and development. Any change in this pattern in recent years seems to be mainly associated with vaccines that are potentially lucrative in the developed world (such as the hepatitis vaccines) rather than vaccines whose main application is in the developing world (such as a vaccine for malaria). This seems to be the case whatever the dimensions of the burden of disease.

For these and other reasons, the IOM committee charged with studying the impediments to U.S. participation in the international Children's Vaccine Initiative (World Health Organization) saw real merit—in fact, necessity—in the establishment of a National Vaccine Authority (NVA). The NVA would have both the capacity and the budget to support vaccine research and to engage in joint ventures with the private sector in the development, manufacture, and clinical trials associated with vaccine development. In this way, some of the development costs would be borne by government, yet the private sector would have the opportunity to realize profits. There have been stirrings of interest in some form of this model within the biotechnology community—and stirrings of opposition in the few large integrated pharmaceutical companies currently engaged in vaccine manufacture.

An alternative strategy might be the establishment of a consortium of pharmaceutical houses that would be encouraged to engage in this area of activity through some form of subsidy. But the subsidy proffered would likely have to be big enough to overcome industry perceptions of a less-than-attractive market. This is not solely a matter of perception: "new-tech" vaccines will not be inexpensive in the early phases of market launch, and it is not likely that developing country markets can demonstrate an ability to pay or levels of demand that would be realistic, never mind attractive, for pharmaceutical companies and their investors. Both the NVA and the consortium strategy would require additional outside capital, whether for vaccine development in the developed world or for creation of production facilities in developing countries.

Indeed, what about vaccine *development* in the developing world? Most efforts to promote local vaccine research, development, and production in developing countries have been flawed where they have not failed outright. This has stemmed in part from an inadequate indigenous technical capacity and in part from what have been, in some instances, massive problems in quality control. There also have been some quite vexing issues surrounding international property rights. All these issues could be addressed in some way by the international development community by making a serious, coordinated attempt at capacity-building and thoughtfully tackling the intellectual property right issues. Any such efforts thus far appear to have been scattered and tormented by a variety of institutional *angsts*.

Expanded Primary Health Care

There is a tendency to think of primary care as somehow nontechnological or, at most, low tech. In fact, integrated systems of health care delivery are explicitly subsumed in the definition of *technology* that has been used by the Institute of Medicine in its fairly extensive body of work on medical technology, its applications, and its assessment. The evolution of primary health care in the developing world after the international meeting at Alma-Ata has provided the structural "shelter" for the development of some rather remarkable technologies. Oral rehydration therapy (ORT), a quite elegant technology firmly rooted in a great deal of basic and applied science, heads that list. Despite that fact, ORT has not been particularly well integrated into protocols for diarrhea case management in the United States.

But this is not surprising given the driving role of "high" technology in U.S. health care. Analogously, the U.S. patient population generally prefers a personal physician over other types of health care provider. Yet it is extremely unlikely that the desire to have a physician in every small town in the United States can ever be achieved. The developing countries have learned this faster and dealt with it more creatively and consistently than Americans have. In fact, as managed care inevitably expands in the United States, so will the use of a variety of health care providers, including nurses, nurse practitioners, physician assistants, technicians of different types, and community workers. This does not imply the absence of technology; it means only that the hands that apply it will not necessarily be those of a physician.

A number of the technologies already being developed in anticipation of this shift in the hierarchy of health care delivery will be nicely transferrable to developing countries. For example, "telemedicine" now allows health care providers in rural communities to communicate with an academic health center so that a patient can be seen, heard, and even partially examined by a consultant at a distance. More futuristically, research under way by medical device manufacturers and the Department of Defense's Advanced Research Projects Agency (ARPA) is seeking to apply virtual reality in ways that will actually permit the surgical procedures themselves to be carried out at a great distance. Indeed, the same fiber-optic techniques that surgeons now use to remove a gall bladder can be employed to transmit images over thousands of miles and actually manipulate medical equipment in distant settings through robotics. But one cannot be naive about the immediate potential of these technologies; they will require not only the installation of the corresponding technology *in situ*, but also its meticulous maintenance, excellent antisepsis, and a quality of patient care commensurate with the intervention and illness in question.

Chronic Diseases

Many developing countries are now undergoing a demographic change in which the population is aging and many of the chronic illnesses associated with developed countries, such as cancer and heart disease, are becoming disconcertingly prevalent. Critical to the management of these illnesses are cost-effective strategies for their prevention and their large burdens of mortality and morbidity.

None of these strategies is more important than informed tobacco policies. In societies in which cholesterol levels are low, cigarette smoking in itself is not a major predisposing factor toward heart disease. As cholesterol levels rise, however, there is a multiplier effect, ranging from two to four, which means that the use of cigarettes multiplies the impact of those rising levels. Thus developing countries that begin to improve their nutrition in a direction that increases serum cholesterol levels also will see rising heart attack rates, which will in turn escalate in the presence of extensive cigarette smoking. Methodologies to minimize smoking and public policy interventions that limit the introduction of cigarettes or their accessibility are therefore critical. But this will not be easy; the behavioral component is central, the incentives to the national and international producers of the raw and finished products are great, and the addiction of nicotine is real.

Some of the technologies that can dramatically improve health are not usually perceived as "health technologies." For example, in many parts of the world unventilated indoor cooking produces emphysema and other chronic lung diseases and, at a minimum, exacerbates the ordinary respiratory infections that are so prevalent in developing countries. As populations age, the magnitude of that accumulated burden can only increase and accelerate. Over the years there have been experiments with such appropriate technologies as low-smoke stoves and improved ventilation arrangements, as well as different energy sources. But apparently no innovations have been well adapted and distributed. This does not mean that the need does not persist or that the returns to health would not be considerable.

Microbial Threats

For developing nations, no development is more ominous than the globalization of illness, particularly infectious diseases. The recent Hanta virus outbreak and the emergence of resistant tuberculosis in the United States in themselves have increased public awareness dramatically.

A prescient report on this topic, produced in 1992 by an Institute of Medicine committee cochaired by Nobel laureate Joshua Lederberg and Robert Shope,[15] has provided the basis for a well-articulated plan by the Centers for Disease Control and Prevention for addressing the multiple facets of this very real problem. Heading the list of requisite actions is the establishment of a global surveillance system and renewal of pertinent research. The issues that need to be addressed range from low to fairly high technology, and this spectrum of needs, some of them urgent, could be addressed usefully in the development and scientific communities.

Perhaps the most urgent issue in the area of emerging infections is the virtual collapse of the antibacterial weapons systems. Not only must the battle to understand individual diseases and ways to combat them be continued, but it is also necessary to address the worldwide threat of the increasing antibiotic resistance of a number of organisms that produce major disease burdens. Because bacteria and many other microbial agents reproduce at very rapid rates, there is a high probability that a single bacterium of the millions of descendants of the original infecting agent will undergo a spontaneous mutation, making it resistant to an antibiotic. The antibiotic may manage to kill all of the other organisms involved in an infectious episode, but the single resistant organism could remain nonetheless and proceed to reproduce in yet another case of the survival of the fittest.

A major strategy for dealing with this situation is the use of multiple drugs. The probability that a mutation that occurs in one in a million cell divisions will produce a genetic resistance to two antibiotics is approximated by the product of that frequency—that is, a million times a million chances. In some types of tuberculosis, use of three drugs is desirable.

This is both a technological and a behavioral matter. One reason antibiotic technology has failed is its overuse by providers and poor adherence to treatment regimens by patients. Designing methodologies by which individuals can take multiple drugs simultaneously and assuring compliance by providers and patients with prescription and case management regimens are major research challenges.

HIV Infection

Hanging over the entire world is the specter of HIV infection because no easy solution is in sight. A vaccine is not on the horizon. Moreover, it is unlikely that a crash investment in the development of such a vaccine will dramatically decrease the time to its availability in the absence of some new intellectual breakthroughs in the understanding of viral variation or some plain luck. The current $1.2 billion NIH budget is probably more than adequate to assure that reasonably promising leads are followed. In the absence of such a vaccine, the very expensive antiviral drugs available have been shown to be of only marginal value.

Modification of behavior continues to be the only hope of substantially decreasing the rate at which the disease spreads. This will require an in-depth understanding of the variations in the cultural, social, and ethical values and mores among the societies in which HIV is prevalent. In the United States, this means the variation among San Francisco, Los Angeles, and New York, all of which have very different scenarios for the spread of infection. The modes of transmission in Africa and in Southeast Asia differ even more. This is an area in which either the technology fails or it is defined inappropriately.

INTEGRATED SCIENCE AND TECHNOLOGY CENTERS

This cursory view of the kinds of technologies needed to advance health in low-income countries has provided little insight into how these technologies might be more speedily and efficiently developed. A system in which most research and development are carried out in industrial countries and are dominated by the U.S. market will not serve developing countries well, if only because differences in patterns and manifestations of disease, standards of living, environment, and available resources mean that research questions asked in the developed world are not always or sufficiently responsive to needs in developing countries.

Increasingly, it has been recognized that creative research efforts require a critical mass of scientists from different disciplines—some pursuing fundamental research and others more clinical applications—but all with access to expensive equipment and laboratories to house it. Regardless of the undeniable improvements in electronic and computer communications, the need for scientists working together to interact daily is unchanged, if only because science is increasingly cross-disciplinary and thus demanding of team effort. This is no less true in developing countries and may be even more urgent in some respects: efforts to develop and extend technologies in most such countries have often failed as a result of the lack of well-articulated teams of trained personnel to facilitate real technology transfer.

Over the years, there has been talk about using the CGIAR (Consultative Group on International Agricultural Research) model for the health sector. That interest has waxed and waned so that, with the exception of a few institutions such as the International Centre for Diarrheal Disease Research in Bangladesh, little has happened. A strategy for the development and continued support of regional integrated science and technology centers—located in key sites around the world, financed through public-private cooperation, and challenged to confront important research and development issues within a given region—is more likely to succeed than a large number of decentralized facilities marginally staffed and equipped.

Such regional centers can serve as critical training sites for local young people who wish to acquire research and technology transfer skills. If the quality of the centers were comparable to that of establishments in the higher-income countries, then scientists from the industrialized as well as the developing world would be motivated to spend significant amounts of time there. The tenure of developing country scientists would be limited to ensure their return to their countries of origin to continue their work and to serve as agents of technology transfer. In some cases, even entire cadres of researchers and scientists would be trained so that when they return to their own countries they could, properly equipped, carry on research and development and the transfer of technology.

Personnel in these centers should include men and women trained to organize and conduct clinical trials, health service researchers able to assess technologies for total effectiveness, as well as social and behavioral scientists able to address prevention, ethics, and equity. Facilities for production of materials for clinical trials should be available at the centers or in connection with them.

While the costs and the political sensitivities and difficulties associated with siting such establishments cannot be underestimated, and the familiar arguments about making the strong stronger at the expense of the weak cannot be ignored, these sensitivities and arguments have too often served as an excuse for inaction. The world is changing too fast and the old models of technology development and transfer have been discredited in too many ways to cling to them. It is time to move on to new ideas and new models.

NOTES

1. In the United States, FDA, the Environmental Protection Agency, and other health agencies play dominant but varying roles in setting the conditions for acceptance of a new technological innovation. That variation stems from differences in the points at which the agency in question enters the evaluative and regulatory process and the ways its involvement affects an innovation's development, adoption, and application.

2. National Science Board, *Science and Engineering Indicators* (Washington, D.C.: U.S. Government Printing Office, 1991), 100.

3. The relationships among population growth, environmental degradation, and poverty have been much debated, notably at the recent UN conferences on environment (Rio de Janeiro) and population (Cairo). The population/poverty/environment "spiral" is well articulated in the recent UNICEF report, *The State of the World's Children* (New York: Oxford University Press, 1994). The accompanying narrative cites better family planning options as a crucial element in arresting the downward momentum of that spiral.

4. Institute of Medicine, *Applications of Biotechnology to Contraceptive Research and Development: New Opportunities for Public-/Private-Sector Collaboration* (Washington, D.C.: National Academy Press, 1995). Funding for this study was provided by the Rockefeller Foundation, Andrew W. Mellon Foundation, National Institutes of Health, and U.S. Agency for International Development.

5. World Bank, *World Development Report 1993: Investing in Health* (New York: Oxford University Press, 1993).

6. Institute of Medicine, *New Vaccine Development: Establishing Priorities,* vol. 1 (Washington, D.C.: National Academy Press, 1985).

7. I. T. Elo and S. H. Preston, "Effects of Early Life Conditions on Adult Mortality: A Review," *Population Index* 58 (1992): 186-212; D. J. Jamison et al., ed., *Disease Control Priorities in Developing Countries* (New York: Oxford University Press, 1993); W. H. Mosley and R. Gray, "Childhood Precursors of Adult Morbidity and Mortality in Developing Countries: Implications for Health Programs," in *The Epidemiological Transition: Policy and Planning Implications for Developing Countries,* ed. J. Gribble and S. H. Preston (Washington, D.C.: National Academy Press, 1993).

8. Office of Technology Assessment, *Identifying Health Technologies that Work: Searching for Evidence* (Washington, D.C.: U.S. Government Printing Office, 1994). (Available from Superintendent of Documents, P.O. Box 37194, Pittsburgh, PA 15250-7974. Stock #052-003-01389-4.)

9. National Research Council and the John F. Kennedy School of Government, Harvard University, *Violence in Urban America: Mobilizing a Response—Summary of a Conference* (Washington, D.C.: National Academy Press, 1994).

10. National Research Council, *Understanding and Preventing Violence* (Washington, D.C.: National Academy Press, 1993).

11. A. Kleinman et al., *World Mental Health: Problems and Priorities in Low-Income Countries* (New York: Oxford University Press, 1995).

12. See R. Kaplan, "The Coming Anarchy," *Atlantic Monthly,* February 1994, 44-75 *passim.*

13. K. R. Mahaffey, "Environmental Lead Toxicity: Nutrition as a Component of Intervention," *Environmental Health Perspectives* 89 (1990): 75-78.

14. See, among other works, *The Lives of A Cell: Notes of a Biology Watcher* (New York: Bantam, 1975).

15. Institute of Medicine, *Emerging Infections: Microbial Threats to Health in the United States* (Washington, D.C.: National Academy Press, 1992).

Sustainable Development:
Mirage or Achievable Goal?

ROBERT M. WHITE
President, National Academy of Engineering

Environmental issues are quintessential global problems that require policy makers to consider all the options offered by science, technology, economics, and social science if they are to address these issues wisely. Policy makers also must ask themselves: Is environmentally sustainable economic growth a mirage or an attainable goal? If such growth is attainable, where can intellectual and financial investments make a substantial difference?

For many years the conventional wisdom, especially in much of the developing world, was that environmental protection and economic development were largely incompatible. In 1987, however, the report of the United Nations' Brundtland Commission, *Our Common Future,*[1] argued against this view. The idea that the environment and development are not antithetical became the philosophical framework for the UN Conference on Environment and Development, which convened in Rio de Janeiro in 1992, and it is now the overarching philosophy guiding world actions. Indeed, the phrase "sustainable development" has become the global environmental watchword, capturing the idea that economic development can be environmentally sustainable. Moreover, this concept suggests that sustainable development not only is a desirable goal, but also is necessary to prevent eventual global, societal, and environmental collapse. Early adherents to this view envisioned a sufficient transfer of resources from the industrialized to developing nations to enable this grand global bargain to be consummated. The Global Environmental Facility (GEF) of the World Bank was the result.

But now society is on the road from Rio to reality, and the road is riddled with potholes—political, economic, technological, scientific, and otherwise. Charting the pathways to sustainable economic growth will require understanding of the forces that lead to unsustainability: population growth, the drive for economic and social equity, the need for adequate food and energy, and the longtime trend toward increased industrialization to provide goods and services.

The complexity of this global dilemma stems from the lack of ways to address the interconnections among these driving forces. A growing population requires more land for human habitation and food production, which leads to soil erosion and the degradation of virgin lands. Animal habitats are affected, which leads in turn to the extinction of some species. The net result is an impoverished resource base to sustain life. Or again, increased industrial and agricultural production to achieve higher living standards requires more energy, thus increasing greenhouse gas emissions and the production of other pollutants. The consequences are climatic warming, urban air pollution, and degraded aquatic systems.

Approaches to environmental problems are rendered even more difficult because many environmental problems are global, requiring action across nations. Yet any action only can be taken locally in countries with different political, social, and economic systems, cultures, levels of education, and capacities in science and technology.

The dilemma is an ancient one. Two hundred years ago, Thomas Malthus pointed out the expected long-term consequences of unrestrained population growth in the face of a limited food supply. In more recent years, studies such as those of the Club of Rome have addressed the consequences of unrestrained growth in the face of limited resources. The Club of Rome's 1972 report *Limits to Growth* foresaw nothing but a global apocalypse.[2]

But the apocalyptic nature of many of these analyses of global systems has so far failed the test of reality. For example, Malthus could not foresee the revolution in food production that science and technology would produce. The green revolution has turned food-importing nations into food-exporting nations, and the future promises continued quantum leaps in food production as genetic engineering yields greater and more resilient crop strains and promises to multiply key aspects of animal productivity. Energy supplies have systematically increased despite predictions that reserves of fossil fuels will decline. Science and technology have made possible the discovery of new energy sources even as they have made nuclear and renewable sources technically practical. Technology itself has been central to the processes of change that have made it possible to avert predictions of environmental calamities by providing expanded options in infrastructure, habitat, and lifestyles for what in the end is determined on socioeconomic grounds through the political process.

The effects of advances in science and technology historically have been hard to predict. For example, the following scientific and technological discoveries and developments took place less than 50 years ago, yet they already have had profound impacts the way we live, think, and conduct business and our everyday lives. The first commercial jet aircraft flew only in 1949, although the first experimental jet was introduced in 1939. The power of the atom, when harnessed, was unfortunately demonstrated in the destruction of Hiroshima and Nagasaki in 1945. In 1953, Watson and Crick unraveled the secret of the double helix and the DNA molecule, opening the era of molecular biology and genetic engineering and technology. Earth-orbiting satellites were not introduced until 1957 by the Russians, and the transistor and its progeny the microchip, the personal computer, and modern communications did not make their debut until 1948. Fiber optics and the laser are only 30 years old. And the birth-control pill arrived in 1957. Any attempts to predict future scientific discoveries or technological developments will be uncertain at best. Indeed, one cannot extrapolate the future assuming a dumb world in which intellectual power and humanity's capacity to choose is straitjacketed. Predictions of the future that assume an unchanging response by society are doomed to apocalyptic conclusions.

Historically, scientific discoveries and technological developments have serendipitously ameliorated environmental deterioration or have produced unanticipated deleterious effects. For example, gas from oil wells was flared as a useless byproduct of oil production until technology provided ways to use it economically. The use of creosote to preserve wooden railroad ties effectively protected forests by reducing the number of trees harvested. The internal combustion engine changed the face of society, rescuing it from the pollution of horse-drawn carriages and exposing it to pollution from auto exhausts. In more recent times, chlorofluorocarbons (CFCs) were introduced as a safety measure in refrigeration systems to replace ammonia. But their very desirable inert character enabled these chemicals to reach the stratosphere (unaffected by lower atmospheric processes), where their decomposition in the presence of sunlight released the chlorine atoms that are thought to trigger the depletion of stratospheric ozone.

Until recently, technological innovations, with some notable exceptions such as the development of sanitary water supply systems, were motivated by economic interests. Their environmental implications, good or bad, anticipated or unanticipated, were considered side effects. Today, by contrast, environmental benefits are often explicit objectives of technological innovation. For example, the remarkable degree to which digital information technologies can control industrial processes is now minimizing effluents and emissions in ways that were not possible earlier. Modern-day engineering design concepts (for automobiles, for example) take into account the reuse and recyclability of products. And material substitution is minimizing environmental problems and dematerializing products. Outstanding examples are the new chemicals developed to replace chlorofluorocarbons in refrigeration systems and as solvents. Finally, modern human reproduction technologies such as the birth-control pill and RU 486 give men and women more control over the size of their families and the spacing of their children. In short, none of the forces causing global environmental unsustainability is immune from the effects of developments in science and technology, although the adoption of various technologies is sometimes painfully slowed by cultural and social practices and the lack of political will.

Environmental technologies, or better technologies for the environment, cover a wide spectrum of engineering activities that embrace, among other things, the technologies for avoiding pollution or other kinds of environmental deterioration; the technologies for monitoring and assessing environmental conditions or the release of pollutants and effluents; the approaches to controlling industrial processes in order to minimize pollutants entering the environment; and the approaches to restoring environmental ecosystems. Markets in the developed world for environmental technologies are large, and export markets in the developing world can be expected to follow in the years ahead. The market today for environmental technologies is about $300 billion a year and may reach $425 billion in a few years.

DEFINING THE PROBLEMS

As World Bank and other reports point out, perhaps the most pressing global environmental problem is the lack of clean water. People in developing countries suffer from a disproportionate amount of water-borne diseases. For example, the United Nations Children's Fund estimates that about 40,000 children die every day, mainly from preventable water-borne diseases. But to solve this problem there is no need to develop new technologies: it has been known for many decades how to devise sanitary water supply systems.

The second most pervasive environmental problem is urban air pollution. Here again much is known about the technologies for controlling this problem, but as populations continue to concentrate in large cities, this problem will only grow worse. In fact, the number of cities with populations of over 10 million is expected to grow from the present 13 to over 25 in the next 15 years. These megacities will give new urgency to the need to address urban air pollution and other urban environmental problems.

At an even more fundamental level is soil erosion. Soil quantity and quality are being rapidly depleted in many countries of the world. As the pressure to increase food production continues, lands that are more marginal will be brought into use. Just as for water sanitation, however, the technologies to improve soil conservation are well known; they only need to be adopted worldwide.

Finally, there are the truly global environmental problems whose causes and effects are widely dispersed and whose resolution will require international action. These problems include but are not restricted to acid deposition, climatic warming, ocean oil spills, and loss of biodiversity.

PROSPECTS FOR THE FUTURE

The present great wave of new technologies and technological concepts collectively represents a new environmental technological offensive. Properly directed and financed, this offensive could open pathways to an environmentally sustainable future as well as restore damaged environments. Technological innovation by itself is a necessary, but insufficient, means to that end. Wise socioeconomic and political choices also must be made as society comes to grips with the inevitable trade-offs between environment, population size and distribution, lifestyle, and economic resources, to form the basis for guiding useful technological development.

Take, for example, the progress that has been made in energy technologies. The production, distribution, and use of energy have widespread, diverse environmental consequences. But advances in energy production, storage, and use now make the entire energy supply and demand system more efficient and less demanding of fossil and other fuels. Combined-cycle gas turbines, new emission-control systems, improved technologies for suppressing auto emissions, increased use of less-polluting fossil fuels such as natural gas, increased use of renewable energy sources, as well as a host of new demand-side energy technologies such as more efficient lighting, appliances, and insulation are conscious attempts to minimize environmental impacts.

What is taking place is encouraging. In fact, the entire industrial approach to producing goods and services is being viewed in a new way as a living system. Just as in biological systems, industrial metabolism is measured by the inputs of energy and resources and by the outputs of useful products and "wastes" of various kinds—emissions to the atmosphere, effluents into rivers, solid wastes into landfills. Useful products also become wastes as soon as they are consumed and discarded. Although some of these discards and wastes can be used in other production processes, others, unfortunately, are widely dispersed and are irretrievably dissipated into the environment.

But one company's or person's waste can be another's valuable input—an industrial analogy to natural ecosystems—and the concept of industrial ecosystems is now taking hold. Natural ecosystems usually are sustainable or slowly changing except for external forces. The primary energy and resource inputs result in a food chain and the behavior patterns of living organisms that sustain themselves and their systems. There are few wastes in natural ecosystems. Wastes are generally inputs to other parts of the system, thereby sustaining the diversity of life and plant forms.

Can technology help humans to mimic such natural systems? Could the wastes in one part of an industrial ecosystem become inputs to other parts of the system? Using technology, researchers should soon come close to providing acceptable systems for social choice. The present crude attempts to mimic natural ecosystems are but the first halting steps toward sustainable economic growth—the environmental Holy Grail. The practices needed are just beginning to be formulated and introduced. Recycling mimics some aspects of natural ecosystems, and a variety of recycling practices are now taking hold. For example, scrap iron has long been recycled to produce iron and steel, and attempts are being made to recycle aluminum, copper, and gold.

The incentives to recycle are largely economic in a free-market system. When free-market economic incentives are lacking, they have been created artificially through legislated regulatory measures or taxes—remedies for market failures that cannot and do not reflect the costs of externalities. Some countries now mandate recycling through "take back" legislation in which the manufacturer takes responsibility for the reuse of product materials at the end of the product's useful life. Recycling of paper, glass, and other kinds of wastes also is now mandated in many communities. Certain mixtures of gasoline and oxidants, such as methanol or ethanol, are mandated as well to reduce auto emissions. In short, it is now possible to choose among economic costs and possibilities in order to elicit the desired environmental results. Even though people frequently lack the political will to accept the trade-offs in cost and lifestyles, technology can help to make these trade-offs more acceptable.

Kalundborg, Denmark, is a particularly pertinent and successful example of the application of industrial ecological principles and the degree to which it is presently possible to mimic natural ecosystems. This small industrial city is home to a Statoil Corporation oil refinery; Denmark's largest power plant, Asnaesverket; a plaster board manufacturing plant, Gyproc; and Novo Nordisk, a biotechnology plant that produces 45 percent of the world's insulin and 50 percent of the world's enzymes. In this city, which is surrounded by a farming community: refinery wastewater is used for power plant cooling; excess refinery gas and sulphur recovered by the refinery is used by Gyproc to produce plaster board; biological sludge from the pharmaceutical plant is used by farmers; steam from the power plant is used by the pharmaceutical company; fly ash from the power plant is used by cement manufacturers in a different town; and waste heat from the power plant is used by the municipality for its heating distribution system, as well as for fish farming. As a result, resource use is reduced: oil by 19,000 tons a year, coal by 30,000 tons a year, and water by 1.2 million tons a year. Emissions are reduced as well: carbon dioxide by 130,000 tons a year and sulphur oxide by 25,000 tons a year. Kalundborg is a very clean industrial town.

Such examples are encouraging, but by themselves they will fall short of the goal of sustainable economic growth. It is in humanity's power, however, by investing its intellectual and financial resources in promising new technologies, to change population growth rates, the modes of food and energy production, and its industrial and agricultural processes. It is also possible to reverse and restore natural environmental systems that through neglect and misuse have deteriorated, notwithstanding the formidable cultural, religious, and social obstacles that must be overcome.

The World Bank and others could invest in both the intellectual framework for advancing the concepts of industrial ecology and the actual demonstration in developing nations of new environmentally sound industrial practices. Also promising as an area of focus for the World Bank is technological capacity-building. Scientists speak frequently of science capacity-building. In fact, the START program proposed by the International Council of Scientific Unions would educate and train scientists in order to create an indigenous understanding of environmental science. And the United States and other Western Hemisphere nations have agreed to an inter-American network of such environmental training centers.

Another fruitful area is in the support of systems studies, not for prediction of the course of future events but to indicate possible areas of research at the interface between the forces that drive unsustainability: population, economics, environment, and technology. An overarching framework is needed for considering the complexity of the issue. Groups around the world that already are considering pathways to sustainability could progress much faster with additional support.

Finally, improved ways of communicating and demonstrating best environmental practices in various industrial sectors are needed, as well as international support for promising environmental technologies.

Sustainability is not a new concept; traditionally it has been applied in the management of renewable resources—for example, fisheries and forestry. It is now time to enhance industry's ability to mimic natural ecosystems, thereby helping today's complex industrial society reach an acceptable level of sustainability.

A vision of the environmental future essential to the survival of humanity is now emerging, and it is within society's power to make the choices and marshal the efforts necessary to travel this road. This is a task for global collaboration, and nothing could be more worthy of humanity than such a crusade. As humanitarian and environmentalist René Dubos has said, "Trends are not destiny."

NOTES

1. World Commission on Environment and Development, *Our Common Future: A Report of the World Commission on Environment and Development* (New York: Oxford University Press, 1987).

2. Donella H. Meadows et al., *The Limits of Growth: A Report for the Club of Rome's Project on the Predicament of Mankind* (London: Earth Island Limited, 1972).

APPENDIX

Symposium Participants

World Bank/National Research Council
Symposium on Marshaling Technology for Development
Arnold and Mabel Beckman Center
Irvine, California
November 28-30, 1994

Bruce Alberts
National Academy of Sciences

Dennis Anderson
Industry and Energy Department
World Bank

Richard E. Balzhiser
Electric Power Research Institute

Jordan J. Baruch
Jordan J. Baruch Associates

James Bond
Industry and Energy Department
World Bank

Carlos Braga
International Economics Department
World Bank

Harvey Brooks
John F. Kennedy School of Government
Harvard University

Annice Brown
Asia Technical Department
World Bank

George Bugliarello
Polytechnic University

John Campbell
Office of International Affairs
National Research Council

Elkyn Chaparro
Finance and Private Sector Development
World Bank

Prita Chathoth
Electronic Media Center
World Bank

Praveen Chaudhari
IBM Thomas J. Watson Research Center

Caroline Clarke
Board on Natural Disasters
National Research Council

E. William Colglazier
National Research Council

Rita Colwell
University of Maryland Biotechnology
Institute

Jose M. Costa
Science, Technology and Education
    Section
Delegation of the Commission of the
    European Union

Carl Dahlman
Private Sector Development Department
World Bank

Reynaldo dela Cruz
National Institute of Biotechnology
University of the Philippines

Augustin del Rio
Vitro Corporativo
Mexico

Neal Dickerson
Electronic Media Center
World Bank

Gerald P. Dinneen
National Academy of Engineering

Donna Gerardi
Coordinating Council for Education
National Research Council

Paul Gilman
Commission on Life Sciences
National Research Council

David Godfrey
Manufacturing Resources, Inc.

Melvin Goldman
Asia Technical Department
World Bank

Michael P. Greene
Office of International Affairs
National Research Council

Demissie Habte
International Centre for Diarrheal Disease
    Research, Bangladesh

Kristin Hallberg
Private Sector Development Department
World Bank

Nagy Hanna
Asia Technical Department
World Bank

Janet Hansen
Board on International Comparative
    Studies in Education
National Research Council

Richard R. Harwood
Michigan State University

Sidney F. Heath III
AT&T

Lee Hoevel
AT&T Global Information Solutions

Lauritz Holm-Nielsen
Education and Social Policy Department
World Bank

Christopher Howson
Board on International Health
Institute of Medicine

Dean Jamison
Latin America and Caribbean Technical
    Department
World Bank

Peter Knight
Electronic Media Center
World Bank

Sabra Bissette Ledent
Consulting Editor
National Research Council

Alan M. Lesgold
Learning Research and Development
    Center
University of Pittsburgh

Dev Mani
Division on Infrastructure, Energy, and
  Environmental Engineering
National Research Council

John S. Mayo
AT&T Bell Laboratories

Alexander McCalla
Agriculture and Natural Resources
  Department
World Bank

Robert McDowell
North Carolina State University

Stephen Merrill
Board on Science, Technology and
  Economic Policy
National Research Council

Lourival Carmo Monaco
Financiadora de Estudos e Projetos
Brazil

Mohan Munasinghe
Environment Department
World Bank

Sharda Naidoo
Council for Scientific and Industrial
  Research
South Africa

Susan Offutt
Board on Agriculture
National Research Council

Geoff Oldham
International Development Research
  Centre
Canada

Vimla L. Patel
Centre for Medical Education
McGill University

Christian Pieri
Agricultural and Natural Resources
  Department
World Bank

David P. Rall
Institute of Medicine

Proctor Reid
National Academy of Engineering

Jean-François Rischard
Finance and Private Sector Development
World Bank

F. Sherwood Rowland
National Academy of Sciences

Robert Schware
Industry and Energy Department
World Bank

Julian Schweitzer
Latin America and Caribbean Region
World Bank

Kenneth I. Shine, M.D.
Institute of Medicine

Evald Emilevich Shpilrain
Institute for High Temperatures
Russian Academy of Sciences

Richard Stern
Industry and Energy Department
World Bank

Eduardo Talero
Industry and Energy Department
World Bank

Paul Uhlir
Commission on Physical Sciences,
  Mathematics, and Applications
National Research Council

Peter Urban
Technical and Environment Department
International Finance Corporation

N. Vaghul
Industrial Credit and Investment
   Corporation of India Limited

Jacques Van Der Gaag
Population, Health and Nutrition
   Department
World Bank

Robert M. White
National Academy of Engineering

Wendy D. White
Office of International Affairs
National Research Council

John D. Woods
Imperial College of Science, Technology
   and Medicine
United Kingdom

James B. Wyngaarden
National Academy of Sciences